WHERE THE WEATHER MEETS THE ROAD

A RESEARCH AGENDA FOR IMPROVING ROAD WEATHER SERVICES

Committee on Weather Research for Surface Transportation:
The Roadway Environment

Board on Atmospheric Sciences and Climate

Division on Earth and Life Studies

Transportation Research Board

NATIONAL RESEARCH COUNCIL
OF THE NATIONAL ACADEMIES

THE NATIONAL ACADEMIES PRESS
Washington, D.C.
www.nap.edu

THE NATIONAL ACADEMIES PRESS 500 Fifth Street, NW Washington, DC 20001

NOTICE: The project that is the subject of this report was approved by the Governing Board of the National Research Council, whose members are drawn from the councils of the National Academy of Sciences, the National Academy of Engineering, and the Institute of Medicine. The members of the committee responsible for the report were chosen for their special competences and with regard for appropriate balance.

This study was supported by Contract No. DTFH61-01-C-00036 between the National Academy of Sciences and the Federal Highway Administration. Any opinions, findings, conclusions, or recommendations expressed in this publication are those of the author(s) and do not necessarily reflect the views of the organizations or agencies that provided support for the project.

International Standard Book Number 0-309-09136-5 (Book)
International Standard Book Number 0-309-52980-8 (PDF)

Library of Congress Catalog Card Number 2004101858

Additional copies of this report are available from the National Academies Press, 500 Fifth Street, N.W., Lockbox 285, Washington, D.C. 20055; (800) 624-6242 or (202) 334-3313 (in the Washington metropolitan area); Internet, http://www.nap.edu

Cover photo: Amanda Staudt, National Research Council.

Printed in the United States of America.

THE NATIONAL ACADEMIES

Advisers to the Nation on Science, Engineering, and Medicine

The **National Academy of Sciences** is a private, nonprofit, self-perpetuating society of distinguished scholars engaged in scientific and engineering research, dedicated to the furtherance of science and technology and to their use for the general welfare. Upon the authority of the charter granted to it by the Congress in 1863, the Academy has a mandate that requires it to advise the federal government on scientific and technical matters. Dr. Bruce M. Alberts is president of the National Academy of Sciences.

The **National Academy of Engineering** was established in 1964, under the charter of the National Academy of Sciences, as a parallel organization of outstanding engineers. It is autonomous in its administration and in the selection of its members, sharing with the National Academy of Sciences the responsibility for advising the federal government. The National Academy of Engineering also sponsors engineering programs aimed at meeting national needs, encourages education and research, and recognizes the superior achievements of engineers. Dr. Wm. A. Wulf is president of the National Academy of Engineering.

The **Institute of Medicine** was established in 1970 by the National Academy of Sciences to secure the services of eminent members of appropriate professions in the examination of policy matters pertaining to the health of the public. The Institute acts under the responsibility given to the National Academy of Sciences by its congressional charter to be an adviser to the federal government and, upon its own initiative, to identify issues of medical care, research, and education. Dr. Harvey V. Fineberg is president of the Institute of Medicine.

The **National Research Council** was organized by the National Academy of Sciences in 1916 to associate the broad community of science and technology with the Academy's purposes of furthering knowledge and advising the federal government. Functioning in accordance with general policies determined by the Academy, the Council has become the principal operating agency of both the National Academy of Sciences and the National Academy of Engineering in providing services to the government, the public, and the scientific and engineering communities. The Council is administered jointly by both Academies and the Institute of Medicine. Dr. Bruce M. Alberts and Dr. Wm. A. Wulf are chair and vice chair, respectively, of the National Research Council.

www.national-academies.org

COMMITTEE ON WEATHER RESEARCH FOR SURFACE TRANSPORTATION: THE ROADWAY ENVIRONMENT

JOHN T. SNOW (*Chair*), University of Oklahoma, Norman
ELIZABETH CARTER, WeatherExtreme, LLC, Kings Beach, California
DENNIS L. CHRISTIANSEN, Texas Transportation Institute, Texas A&M University, College Station
BRADLEY R. COLMAN, National Weather Service, Seattle, Washington
PAUL J. DELANNOY, Environment Canada, Gatineau, Québec
FRANCIS B. FRANCOIS, American Association of State Highway and Transportation Officials (retired), Bowie, Maryland
GEORGE L. FREDERICK, Vaisala, Inc., Boulder, Colorado
FRANCES C. HOLT, NOAA's National Environmental Satellite, Data, and Information Service, Silver Spring, Maryland
MARGARET A. LEMONE, National Center for Atmospheric Research, Boulder, Colorado
CURT PAPE, Minnesota Department of Transportation, St. Paul
LELAND SMITHSON, American Association of State Highway and Transportation Officials, Ames, Iowa
DANIEL STOCK*, American Transportation Research Institute, Wakefield, Rhode Island
RICHARD WAGONER, National Center for Atmospheric Research, Boulder, Colorado

NRC Staff

AMANDA STAUDT, Study Director
STEPHEN GODWIN, Director, Studies & Information Services Division, Transportation Research Board
JULIE DEMUTH, Research Associate
PAMELA SZATANEK, Intern
ROB GREENWAY, Project Assistant

*Ending April 2003

BOARD ON ATMOSPHERIC SCIENCES AND CLIMATE

ERIC J. BARRON* (*Chair*), Pennsylvania State University, University Park
ROBERT J. SERAFIN (*Chair*†), National Center for Atmospheric Research, Boulder, Colorado
FREDERICK R. ANDERSON, Cadwalader, Wickersham, and Taft, Washington, D.C.
RAYMOND J. BAN*, The Weather Channel, Inc., Atlanta, Georgia
ROBERT C. BEARDSLEY, Woods Hole Oceanographic Institution, Massachusetts
ROSINA M. BIERBAUM, University of Michigan, Ann Arbor
RAFAEL L. BRAS, Massachusetts Institute of Technology, Cambridge
MARY ANNE CARROLL, University of Michigan, Ann Arbor
KERRY A. EMANUEL, Massachusetts Institute of Technology, Cambridge
CASSANDRA G. FESEN, Dartmouth College, Hanover, New Hampshire
MARGARET A. LEMONE*, National Center for Atmospheric Research, Boulder, Colorado
JENNIFER A. LOGAN, Harvard University, Cambridge, Massachusetts
MARIO J. MOLINA*, Massachusetts Institute of Technology, Cambridge
WILLIAM J. RANDEL, National Center for Atmospheric Research, Boulder, Colorado
ROGER M. WAKIMOTO, University of California, Los Angeles
JOHN C. WYNGAARD, Pennsylvania State University, University Park

Ex Officio Members

ANTONIO J. BUSALACCHI, JR., University of Maryland, College Park
ERIC F. WOOD, Princeton University, New Jersey

NRC Staff

CHRIS ELFRING, Director
SHELDON DROBOT, Program Officer
AMANDA STAUDT, Program Officer
JULIE DEMUTH, Research Associate
ELIZABETH A. GALINIS, Project Assistant
ROB GREENWAY, Project Assistant
DIANE GUSTAFSON, Administrative Associate
ROBIN MORRIS, Financial Officer

---

*Ending December 2003
†Beginning January 2004

Preface

The Federal Highway Administration asked the National Research Council to examine what needs to be done from the research, development, and technology transfer perspectives to improve the production and delivery of weather-related information for the nation's roadways (see Appendix A for the full statement of task). In response the Committee on Weather Research for Surface Transportation was formed. This committee was asked to investigate the current state of knowledge regarding road weather conditions, identify key areas where future strategic investments in research and development would enhance understanding of the road environment, and suggest how knowledge of that environment might be translated into operational production of information for use by decision makers: from drivers, to dispatchers, to traffic managers, to maintainers. In this report the committee provides a framework to engage the transportation and weather communities, as well as other stakeholders, in advancing road weather research and development in the United States.

The committee heard from a wide variety of experts during the study. It soon became apparent that the road weather problem is highly interdisciplinary, spanning micrometeorology, numerical weather prediction, vehicle technology, communications technology, meteorological and pavement instrumentation, roadway construction and maintenance, routine traffic and emergency management, human factors, and technology transfer. Hearing from experts in all these disciplines provided the committee with a unique perspective of the challenges in addressing the road weather problem and opportunities for advancement. The committee thanks the following speakers who shared their knowledge with the committee: Edward E. Adams, Montana State University; Bob Baron, Baron Services, Inc.; Stanley G. Benjamin, Forecast Systems Laboratory; S. Edward Boselly, Washington State Department of Transportation; Ray Derr, Transportation Research Board; Ian Ferrell, Microsoft Corporation; Robert Hallowell, Massachusetts Institute of Technology Lincoln Laboratories; Jonathan M. Hankey, Virginia Tech Transportation Institute; Eldon L. Jacobson, Washington State Department of Transportation; Henry Lieu, Federal Highway Administration; Jonathan Lister, Vaisala; Clifford F. Mass, University of Washington; Ronald Miller, Ford Motor Company; Joseph Perrin, Jr., University of Utah; Paul Pisano, Federal Highway Administration; Shelley J. Row, Federal Highway Administration; Larry Senn, Washington State Department of Transportation; Harold Smith, Iowa State University; Timothy Spangler, Cooperative Program for Operational Meteorology, Education and Training; Richard C. Steed, University of Washington; Andrew D. Stern, Mitretek Systems, Inc.; Jielun Sun, National Center for Atmospheric Research; Jerry R. Waldman, Surface Systems, Inc.; Samuel P. Williamson, Federal Coordinator for Meteorological Services and Supporting Research; and Chester G. Wilmot, Louisiana State University.

This report provides the committee's consensus findings and recommendations. In Chapter 1 the road weather problem is described. Chapter 2 lays out the committee's vision for the road weather system of the future. In the committee's view this vision could be readily achieved by the year 2020 if sufficient resources were dedicated to furthering road weather research and development, and to fostering the subsequent implementation of those research results. In Chapter 3 the current status of relevant research and development activities in the meteorological and transportation communities is described. It is apparent to the committee that the two diverse communities are not yet working together to address the road weather problem, and that special efforts are needed to bring these two communities together in an effective working relationship. A number of ways to improve the integration of weather and transportation research could yield a substantial payoff, and these along with other opportunities for improving road weather information are identified in Chapter 4. In Chapter 5 the committee recommends how to build an improved road weather program. In Chapter 6 the committee offers some closing thoughts on how developing improved road weather capabilities will increase the nation's capacity to respond more effectively to other large-scale trends.

John Snow
Chair

Acknowledgments

This report has been reviewed in draft form by individuals chosen for their diverse perspectives and technical expertise, in accordance with procedures approved by the National Research Council's Report Review Committee. The purpose of this independent review is to provide candid and critical comments that will assist the institution in making its published report as sound as possible and to ensure that the report meets institutional standards for objectivity, evidence, and responsiveness to the study charge. The review comments and draft manuscript remain confidential to protect the integrity of the deliberative process. We wish to thank the following individuals for their review of this report:

S. Edward Boselly, Washington State Department of Transportation, Olympia
John Campbell, Battelle, Seattle, Washington
Pete Costello, PBS&J, Orlando, Florida
Leon Osborne, University of North Dakota, Grand Forks
Roger Pielke, Jr., University of Colorado, Boulder
Shelley J. Row, Institute of Transportation Engineers, Washington, D.C.
Phil Tarnoff, University of Maryland, College Park
Jerry Waldman, Surface Systems, Inc., Castle Rock, Colorado
Randy Welch, United Parcel Service of America, Atlanta, Georgia
Brian Wolshon, Louisiana State University, Baton Rouge
Da-Lin Zhang, University of Maryland, College Park

Although the reviewers listed above have provided constructive comments and suggestions, they were not asked to endorse the report's conclusions or recommendations, nor did they see the final draft of the report before its release. The review of this report was overseen by Eugene Rasmusson, University of Maryland. Appointed by the National Research

Council, he was responsible for making certain that an independent examination of this report was carried out in accordance with institutional procedures and that all review comments were carefully considered. Responsibility for the final content of this report rests entirely with the authoring committee and the institution.

Contents

Executive Summary

Weather significantly affects the safety and capacity of the nation's roadways. Adverse weather is associated with over 1.5 million vehicular accidents each year, accounting for approximately 800,000 injuries and 7,000 fatalities. Poor road or visibility conditions often cause drivers to slow down, thereby substantially reducing roadway capacity, increasing travel times, and in some cases contributing to chain-reaction accidents. It is estimated that drivers endure over 500 million hours of delay annually on the nation's highways and principle arterial roads because of fog, snow, and ice. This conservative estimate does not account for considerable delay due to rain and wet pavement. Because little new highway capacity will be built in the coming decades and vehicle miles traveled are projected to continue increasing, it is likely that maintaining the safety and capacity of the nation's roadways will continue to be a challenge in the foreseeable future.

An improved strategy for addressing the impacts of weather on surface transportation has the potential to help mitigate roadway congestion and save lives. High-quality weather observations and forecasts specific to the roadway environment could help users, including drivers, fleet dispatchers, and law enforcement and emergency management personnel, make better decisions, thereby increasing travel efficiency and safety during adverse weather conditions. Improved road weather information could also help those who construct, operate, and maintain the roadways to better respond to weather problems. More globally, addressing road weather issues could help position the nation's road transportation system to respond appropriately to anticipated major changes in demographics and technology of the twenty-first century.

Despite the advantages of providing enhanced weather information to multiple users, such services have yet to be widely implemented. This reflects a lack of coordination of existing resources and knowledge, as well as a historical shortage of research and development efforts focused on road weather. Further, the development of improved capabilities to provide road weather information has been hindered by a lack of interaction between meteorologists and transportation researchers as well as insufficient funding specifically directed at the road weather problem.

Recent and anticipated advances make the field of road weather ripe for significant progress in understanding and capability. Accompanying advances in meteorology and transportation are improvements in communications, computational capabilities, and geographic information systems, all of which have clear applications to the road weather problem. The notion of smart vehicles in constant communication with weather information providers and traffic control centers, commercial fleets constantly adjusting their routing to avoid anticipated storms, and road maintenance personnel being guided continuously by telemetered in-road sensors no longer needs to be limited to the realm of science fiction.

The Committee on Weather Research for Surface Transportation was formed at the request of the Federal Highway Administration to investigate the current state of knowledge regarding road weather conditions and to recommend key areas of research to enhance operational production of weather-related information for roads (see Appendix A for the full statement of task). The committee's findings and recommendations are designed to provide a framework to engage the transportation and weather communities, along with other stakeholders, to help shape and guide a focused road weather research program. The recommendations will help the weather and transportation research and operations communities capitalize on existing capabilities and take advantage of opportunities for advances. They include a suite of research activities as well as efforts to foster the implementation of an operational road weather capability.

OVERARCHING RECOMMENDATION

1. Establish a focused, coordinated national road weather research program.

The committee finds that there are substantial research questions and opportunities in road weather that warrant a long-term national commitment and therefore recommends the establishment of a focused, coordinated national road weather research program. Sufficient knowledge and experience exist today to initiate such a program; however, some aspects of the program will require additional research and experience before they can be completely defined and implemented. A road weather research program is timely in that it can take advantage of investments being made in weather and transportation research and infrastructure. An incremental investment in integrating and extending these efforts will reap substantial benefits by producing a national road weather information system as part of the nation's emerging infostructure—a network of data collection and dissemination necessary to support real-time management and operation of the roadway transportation system.

Goals of the Program

The goals of the recommended national road weather research program should be to (1) maximize the use of available road weather information and technologies; (2) expand road weather research and development to enhance roadway safety, capacity, and efficiency while minimizing environmental impacts; and (3) effectively implement new scientific and technological advances.

Program Implementation

These goals should be attained by establishing regional centers, national demonstration corridors, and a nationwide solicitation to support individual investigator-led research projects. It is essential that the program foster close, continuing interaction between the research community and those who will use the research results to ensure that users' needs are well integrated into all stages of the research and development process. Indeed, the private sector can serve as an important conduit for transferring research results to operations. In all cases the program must be guided by careful cost-benefit analyses.

The committee recommends regional research centers to develop new technologies, foster technology implementation on regional roadways, and facilitate interaction among federal, state, and local governments, the private sector, and academia. Such centers could be an attractive alternative to establishing individual state programs because many states in a particular region share the same road weather challenges. These regional centers should be interdisciplinary, incorporating weather and transportation researchers as well as relevant practitioners in the public and private sectors. It is essential that the centers be funded competitively with due consideration given to reflect different weather conditions and transportation problems common to their regions. In addition to addressing the needs of their regions, individual centers should specialize to avoid redundant research on the road weather challenges that are multiregional or even national in scale. For example, one center could specialize in fog as a visibility impediment (a national problem), while another could specialize in evacuation strategies (a problem affecting hurricane-prone regions).

Because regional research centers may require substantial investment in infrastructure before research can be conducted, the center selection should consider other collocated resources, such as computing centers or strong university research programs, that could be leveraged. An alternative to regional centers that avoids some of these initial costs is technology development teams, for which researchers are selected competitively from various research centers to work together on a specific research and development problem. This concept has been used successfully by the Federal Aviation Administration's Aviation Weather Research Program to provide efficient and flexible operationally relevant research results. If the technology development team approach is chosen, then strong and effective linkages to the private sector need to be built into the team structure.

The committee also recommends national corridors along two U.S. interstate highways, one running north-south and one running east-west, to demonstrate the effectiveness of various road weather improvements, facilitate nationwide implementation of research results, and provide a seamless stream of road weather information to users. Ideally, the selected highways would traverse several climate zones containing varied land cover, use, and terrain. Further, they would pass through several states that are early adopters of Intelligent Transportation System (ITS) technology and are agreeable to serving as leading states in the implementation of the national road weather research program. Special programs should be established for the federal government and the private sector to partner with these states to showcase road weather products and services. Many of the recommendations made in this report for new technology should be implemented first along these corridors.

Program Management and Funding

Transportation research and management are currently highly decentralized, largely implemented by states, and meteorological research and operations are spread across several federal agencies, universities, and research centers. The private sector has provided many of the targeted road weather services to date. Yet, for the proposed road weather research program to succeed, centralized leadership at the federal level is essential in setting research priorities, administering grants, providing a central repository for research findings, ensuring accountability, and fostering the transfer of new knowledge and technology to operational practice. The Federal Aviation Administration's Aviation Weather Research Program provides a good model for the federal leadership needed for a road weather research program. The committee recommends that the Federal Highway Administration have the leading role in the national road weather research program and that the National Oceanic and Atmospheric Administration be a lead partner in this effort. The federal government should establish an interagency coordinating council to guide the national road weather research program. In addition to the Federal Highway Administration, participation on this council should include at a minimum the National Oceanic and Atmospheric Administration and the National Science Foundation, especially through their joint U.S. Weather Research Program. It is no less critical that the leadership of the proposed road weather research program foster effective partnerships with the private sector in a manner that provides clear processes for interaction.

The committee recommends that long-term, dedicated funding for road weather research sufficient to achieve the program's goals be established as *new* funding in the Federal Highway Administration. The committee estimates that the research program will require a minimum of $25 million per year, based on an assessment of unmet needs, the capacity of the research community to conduct the program effectively, and costs for comparable transportation research initiatives in other areas, such as the Aviation Weather Research Program. The committee believes that many of the research and development challenges that it has identified will require more than a decade to solve; thus a long-term research and development program with appropriate funding commitments is required. Although the Federal Highway Administration is recommended as the primary funding recipient, the committee feels that the dedicated funds for this program will be highly leveraged with funds focused on other initiatives that have an affinity with road weather research and development.

Steps should also be taken to foster effective public-private-academic partnerships in road weather research and technology implementation. Essential partners in this effort include the Federal Highway Administration, the National Oceanic and Atmospheric Administration, the National Science Foundation, the American Association of State Highway and Transportation Officials (AASHTO), academia, state and local governments, the private sector, and nongovernmental organizations such as ITS America and the American Meteorological Society. The committee believes that the preceding recommendations concerning regional research centers, a focus on national demonstration corridors, and widely accessible databases will facilitate such partnerships. Indeed, the contributions from all sectors are essential for the implementation of the program recommended here.

Research Framework

The proposed road weather research program should support research and development in the following five areas:

1. a robust, integrated observational network and data management system specifically designed to meet the need for enhanced road weather research and operational capabilities;
2. a coordinated research effort to increase understanding of road weather phenomena and options for increasing safety, mobility, and efficiency of the nation's roadways during all types of weather;
3. improved modeling capabilities and forecast tools designed to provide relevant, useful information to those who build, manage, maintain, and use the nation's roadways;
4. multiple mechanisms for communicating road weather information to varied users in ways that support better informed decision making; and
5. an infostructure that takes advantage of new technologies to monitor and predict road weather conditions and then effectively convey road weather information to end users.

The committee identified many opportunities for improving understanding, capabilities, and products in these five areas. In each of these research foci the proposed program should support activities that apply currently available knowledge and tools to the road weather problem, are new research and development efforts, and enable effective implementation of research results. The recommendations that follow highlight key activities in each of these categories.

RECOMMENDATIONS TO OPTIMIZE USE OF AVAILABLE ROAD WEATHER INFORMATION AND TECHNOLOGIES

2. Make better use of currently available road weather information and technologies to increase capabilities for transportation research.

Archive and mine operational traffic observations to assess weather impacts.

Some states and cities monitor traffic throughout their area and routinely collect data from traffic counters, video cameras, and other sensors. Often these data are used for real-time management of the transportation system but are not quality assured or archived in formats or in locations readily accessible to the research community. The committee recommends that the road weather research program take better advantage of these data because they provide many opportunities to learn about impacts of weather on traffic flow and refine estimates of traffic simulation parameters. This will require finding cost-effective ways to quality assure these data, archive them in easy-to-use formats, and make them accessible to researchers.

Integrate available weather information into traffic planning and management models.

The committee recommends research to improve the way weather effects are included in traffic and emergency management models, which currently are used for offline design and planning of transportation systems. Better accounting for weather in these models will allow researchers to assess the benefits of improved weather monitoring and forecasts on transportation operations and enable practitioners to more efficiently manage the nation's roads.

Use real-time weather information in the operation of the transportation system.

The road weather research program should support development of operational modeling systems for roadway management that include real-time weather information. The use of real-time weather information to support the operation of the transportation system is in its infancy but should be encouraged; for example, the road maintenance community has made advances in incorporating real-time weather information in decision support systems for winter operations. Real-time road weather information would also be useful for warning drivers of dangerous weather conditions and for optimizing commercial vehicle operations.

3. Establish a nationwide real-time road weather observation system.

Take advantage of existing observation networks and databases.

The committee recommends that the road weather research program take full advantage of established environmental monitoring networks, including those from in situ and remote-sensing platforms and existing databases on soil type and land use characteristics. As a first step, efforts are needed to integrate and use information from Environmental Sensor Stations and meteorological mesonets with particular attention to seamlessness across state boundaries. A nationwide repository in which weather observations relevant to the roadways are collected and shared by all interested parties would be of great use to the road weather community.

The committee recommends that the road weather research program support the design and establishment of a portal that allows access to (1) weather data relevant to the roadway; (2) state databases of relevant road characteristics, including pavement or concrete composition, road surface depth, and underlying soil type; (3) real-time road surface condition information; and (4) traffic information. Access to these data currently is constrained by the limited data availability and disparity of data formats used by states and municipalities. Having access to these data in standard formats would allow private sector companies to develop forecasting tools that could be easily and reliably applied in different states.

Improve the existing observation system.

To attain an improved road weather observation network the road weather research program should support efforts to establish standards for observing procedures and station siting to ensure that (1) the data are representative of the roadway environment and (2) new road weather stations and traffic monitors are added to optimally address both modeling and special observing needs. The National Weather Service, states, and other entities that perform weather observations should use these standards as they establish new sites to fill gaps in the current observational network.

In addition to improving fixed road weather observation sites, expanded use should be made of mobile observing platforms. Such platforms could consist of a fleet of state- or city-owned vehicles, commercial trucks and buses, and even private vehicles. Vehicle observations of roadway temperature and surface traction could be of great use in providing a more complete and accurate picture of roadway conditions. The road weather community should collaborate with the automobile industry in the design and deployment of these mobile observing platforms and ensure that these data remain open to access.

Maximize utility and quality of road weather information.

The committee recommends that the road weather research program take steps to maximize the utility and quality of road weather information. First, to facilitate sharing among multiple entities, the road weather community should help standardize formats of geolocated data and associated metadata (e.g., the National Transportation Communications for Intelligent Transportation System Protocol Object Definitions for Environmental Sensor Stations). Second, the community should standardize the presentation of road weather information (e.g., in terms of units, colors, or symbols) to user groups across North America. Third, to facilitate integration of information from different research communities, the community should use geographic information systems to analyze, manipulate, and display road weather observations, model output, and forecasts. Fourth, the community should establish standards and implement data quality assurance and control procedures. It is critical to ensure that all road weather observations are of sufficient quality to be used in the numerous desired data applications.

RECOMMENDATIONS TO EXPAND ROAD WEATHER RESEARCH AND DEVELOPMENT

4. Improve weather and transportation modeling capabilities.

Improve accuracy and resolution of road weather forecast products supporting both tactical and strategic decision making.

The road weather research program should support development of tailored road weather information products to support real-time operations and both near-term tactical (0 to 6 hours) and longer-term strategic (> 6 hours) decisions by road users and managers. More accurate weather information products should be developed on fine spatial and temporal scales appropriate to the roadway environment, in part by taking advantage of advances in nested and ensemble modeling approaches. Producing road weather forecasts for the 2- to 6-hour timeframe presents the greatest research challenge because current data become less useful and some models may not have reached the point where they can provide useful output. As a result, this short-term forecast problem likely will remain a focus of the human forecaster and newly emerging statistical tools for the near future. The newly developed National Digital Forecast Database from the National Weather Service provides forecasts across the United States on grid spacing no greater than 5 km and may provide a logical starting point for either user-developed applications or further enhancement by the private sector.

Improve prediction and warning of weather-influenced hazards that rapidly impede roadway use.

The road weather research program should support research to improve predictions and warnings of weather-induced hazards that rapidly impede roadway use, such as flash floods, avalanches, mudslides, and other debris flows. Better understanding of the causes of these hazards and how best to respond to them could help avoid loss of life and property. Existing observational and forecasting capabilities need to be assessed and significant gaps should be addressed.

Improve empirical and numerical modeling techniques to account for both the roadway atmosphere interface and the surrounding environment.

The roadway environment consists of two components: (1) the road surface and (2) the weather and visibility conditions directly above the road. For the former the committee recommends that research be pursued to improve one-dimensional modeling techniques for the roadway-atmosphere interface, the region that extends a few meters above and below the road surface. Advances in this area would enable better forecasts of road surface temperature, leading to better forecasts of frost and black ice. For the latter, empirical and multidimensional modeling approaches should be developed to account for local terrain, land cover, surface composition, and other geographic features that affect the exposure of the roadway to infrared and solar radiation (direct and indirect), wind, and precipitation. Such models need to provide improved forecasts of road surface temperature and condition (e.g., dry, wet, frozen surfaces) as well as fog, wind, and other localized atmospheric phenomena.

Develop end-to-end models that assess and predict weather impacts on roadway conditions and operations.

The road weather research program should support a long-term, interdisciplinary effort to develop end-to-end modeling systems. The required modeling system would have to incorporate (1) current weather and pavement conditions; (2) "future" and forecast weather conditions; (3) a model of road conditions, especially road temperature and traction; and (4) a traffic simulation model that includes the reaction of drivers to such conditions as precipitation, high winds, low visibility, slick pavement, and congestion mitigation strategies (e.g., timing traffic lights, commuter highways, weather-controlled dynamic speed limit signs). Used in an offline planning mode, such a system has enormous potential to enable road managers and emergency responders to develop better proactive responses to extreme weather conditions. As understanding, modeling sophistication, and computer capabilities allow, such models ultimately could be run in real time to assist in the routine management of the transportation system, thereby enhancing safety, capacity, and traffic flow.

5. Develop observing capabilities to measure the performance of road weather forecasts.

The committee recommends that the road weather research program develop special observing facilities to validate forecast tools and verify road weather forecasts. Ideally, these observing facilities would be located along the national corridors and at the regional centers identified previously. In some cases these facilities could consist of permanently instrumented arrays, but to be cost effective most should consist of supporting infrastructure that facilitates the rapid deployment of instrumentation for occasional intensive observation periods. An important aspect of the recommended work would be the collection of micrometeorological measurements necessary to develop, improve, and validate the models.

6. Develop methods for estimating and conveying the degree of confidence in road weather information.

The road weather research program should support the development of means for estimating the inherent uncertainty in road weather information. Advanced weather decision support systems that assimilate many diverse sources of information need to estimate uncertainty and provide easily understood metrics to aid users in making optimal decisions. New methods are needed to enable more effective communication of uncertainty to a wide variety of users.

7. Improve road weather instrumentation.

The road weather research program should support efforts to establish science-based standards for road weather sensors and to implement procedures to test, compare, and certify that sensors meet these standards. The comparison and testing procedures need to ensure that the results are applicable to a field setting. The committee recommends support for the development of improved sensor technology specific to road weather applications, such as for the amount of ice, all-season precipitation, present weather, and chemical composition of the slurry on the road surface. Research is needed on design and deployment of sensor systems to provide data more generally representative of the traveled road surface and that are reliable under a wide variety of weather and traffic conditions.

8. Support human factors research to obtain desired responses to road weather information.

The road weather research program should support research into the human and cultural factors involved in driver decision making under a variety of weather conditions and in driver response to new technologies for the delivery of road weather information. Important aspects of this research include the information, content, and format for optimum response, and safe methods for communicating with vehicle operators, especially during hazardous weather conditions. The Federal Highway Administration's program for an Advanced Traveler Information System has acknowledged that although technology exists to display information to the driving public, there is little knowledge to guide the proper selection of methods and displays. This research should build on the work done by the Advanced Traveler Information System, Commercial Vehicle Operations, and Advanced Traffic Management Systems, and be compliant with the National ITS Architecture Maintenance and Construction Operations user service bundle. Additional research is needed on effective communication of road weather information to traffic managers, maintenance personnel, and emergency managers because they have specialized needs that are significantly different from those of the general public.

RECOMMENDATIONS TO EFFECTIVELY IMPLEMENT NEW SCIENTIFIC AND TECHNOLOGICAL ADVANCES

9. Develop new means to effectively communicate road weather information to a wide range of users.

The road weather community should take advantage of new and upcoming advances in technology that will enhance communication to users. Along with expanded observational capabilities, advanced communications technology must be deployed to enable instruments to transmit data back to central processing centers and to relay processed weather information back to drivers and other users. Advances in wireless communication hold much promise for providing location-specific weather information to mobile users, but the challenge of determining how best to use these communication technologies remains.

10. Pursue nationwide operational capability as research results become available.

Develop a robust national roadway infostructure.

Major components of a roadway infostructure are being developed. The committee recommends that the road weather research program proactively participate in this effort to ensure that the road infostructure of the future incorporates a sophisticated network of road weather observations, including sensors embedded in the pavement, weather stations adjacent to the roadway, water level sensors near flood-prone routes, remote observations from satellite platforms, and instruments on vehicles themselves.

Enable efficient technology transfer.

The committee recommends that the national road weather research program involve operational users as part of the research and development effort from the beginning. The user community for road weather information is broad and includes the general public; commercial shipping and bus transit companies; emergency managers; those who build, maintain, and operate the nation's roadways; and vehicle manufacturers. Early user involvement helps to ensure that the technology transfer path will be efficient and expeditious and will produce results relevant to the needs of the user community.

Improve education and training of road weather information users.

The road weather research program should incorporate training on the use of road weather information and technology into driver education nationwide. This training needs to start with new-driver education and be reinforced periodically as new technology is introduced or other circumstances arise, such as when a driver moves to a new region with a different climate. In commercial settings continuing education on road weather information is recommended for drivers and dispatchers. Special efforts are needed in the education and training for such groups as road maintainers, emergency managers, and traffic managers. When new products or decision support tools are deployed, educational and outreach programs, pilot projects, operational demonstrations, and "table-top exercises" should be conducted on a regional basis.

Seek out synergies and efficiencies between road weather research and parallel efforts regarding other modes of transportation.

The committee recommends that the road weather research program routinely monitor research parallel to surface operations in aviation, rail, and transit to exploit applicable overlap. Many aspects of the proposed road weather research program could benefit research in aviation, rail, and other modes of transportation. Indeed, there are many synergies and efficiencies to be gained by coordinating research on meteorological phenomena that affect all modes and on the development of decision support resources. In particular the well-established aviation weather research program offers many lessons learned that should be examined by the emerging road weather research program.

1

Introduction

Weather has broad and significant effects on the roadway environment. Snow, rain, fog, ice, freezing rain, and other weather conditions can impair the ability of drivers to operate their vehicles safely, significantly reduce roadway capacity, and dramatically increase travel times. Multiple roadway activities, from roadway maintenance and construction to shipping, transit, and police operations, are directly affected by inclement weather. There are substantial costs in loss of life, injuries, and property damage due to weather-related accidents. State and local transportation agencies responsible for maintaining the nation's roadways incur significant costs both directly in responding to phenomena such as ice and snow that impede road operations and indirectly through lost tax revenue if commerce is interrupted. Businesses may be disrupted, especially if they rely on "just in time" arrival of materials and finished goods. There can also be adverse environmental impacts if road treatment chemicals migrate from the road surface to the surrounding environment. Its omnipresence and influence on surface transportation makes weather information a necessity to most roadway users; according to a recent U.S. Department of Transportation survey, "Nearly all drivers surveyed want appropriate, relevant weather conditions included with their traffic information" (USDOT, 2000).

Some road weather information is available to users currently; for instance, there have been important research and development efforts to create better decision support tools for road managers, particularly with regard to snow and ice control activities. Nonetheless a disconnect remains between current research and operations, and additional research could yield important safety and economic improvements for roadway users. Meteorology, roadway technology, and vehicle systems have evolved to the point where users could be provided with better road weather information through modern information technologies. The combination of these technologies has the potential to significantly increase the efficiency of roadway operations, road capacity, and road safety. This report provides a roadmap for moving these concepts to reality.

METEOROLOGICAL CONDITIONS THAT INFLUENCE THE ROADWAY ENVIRONMENT

Nearly all weather conditions affect the roadway environment in some way, typically by affecting visibility, surface traction, or maneuverability of vehicles. Winter weather conditions, including snow accumulation, freezing rain, icy surfaces, and blowing snow, have received the most attention by the transportation community because they can significantly impair the operability of the roadway system over a large region (Figure 1-1). In 2001 110,072 crashes occurred on snowy or slushy roads; consequently, over 1,100 people were killed and nearly 95,000 people were injured. An additional 183,377 crashes occurred in 2001 during snowfall or sleet; nearly

FIGURE 1-1 Traffic on a snowy day. SOURCE: Curt Pape, Minnesota Department of Transportation.

FIGURE 1-2 Car traveling on a rainy night. SOURCE: Amanda Staudt, National Research Council.

790 people were killed and 62,000 people were injured as a result (Goodwin, 2003b). Such conditions occur routinely in the northern states and occasionally in the mid and southern tier states, where their occasional nature makes them more hazardous due to less experienced drivers and lack of adequate winter road maintenance equipment. Roadway maintenance personnel typically respond to winter weather conditions through a combination of plowing and chemical treatments intended to lower the freezing point of water, but challenges still remain in terms of determining the optimal response strategy for a particular winter storm.

Although the impact of winter storms on the roadway system can be quite dramatic, more fatal accidents are attributed to rainy conditions than to any other inclement meteorological condition (Figure 1-2). Over 1.1 million crashes in 2001 occurred on wet pavement, resulting in nearly 5,400 deaths and 511,000 injuries. An additional 688,304 crashes in 2001 occurred when it was raining, resulting in over 3,200 deaths and nearly

309,000 injuries (Goodwin, 2003b). This is not because rain is more dangerous than other phenomena but rather because rain is more common and is not as much a deterrent to road travel as snow or freezing rain. Safety risks include lane submersion, mechanical loss of vehicle maneuverability and control, visibility reduction, and distraction by hail or debris carried by winds. Convective weather, although often on a smaller spatial scale, can present additional hazards, such as lightning, strong winds, and hail. Hail is particularly damaging to vehicles and can seriously distract and stress drivers.

Another common way that weather affects the roadway environment is by causing reductions in visibility due to fog, solar glare, blowing dust or snow, precipitation (and associated vehicle spray), and smoke. In 2001 fog was associated with 43,792 crashes; over 670 people were killed and more than 19,000 people were injured in these crashes (Goodwin, 2003b). Reduction in visibility below a quarter mile poses a threat to driver safety by hampering the ability to see and be seen within a safe reaction distance. When encountered unexpectedly while traveling at high speed, such hazardous conditions can cause serious driver disorientation and lead to dramatic multivehicle accidents. Decreased visibility also can retard traffic flow as drivers properly reduce their speed, thereby reducing the carrying capacity of roadways; however, different drivers will reduce their speed differently in reaction to a sudden reduction in visibility, leading to large speed differentials and increased potential for accidents (OFCM, 2002b).

Winds greater than 25 mph can inhibit the maneuverability and stability of high-profile vehicles, including recreational vehicles, trucks, and buses (OFCM, 2002b). Stronger winds can even topple high-profile vehicles, particularly when they are traveling empty. Wind also reduces visibility through blowing snow, dust, or sand. Conversely, just the right amount of very light wind is required for radiation fog to occur on clear nights. The forecasting of wind extremes, strong and light, is therefore critically important for road safety. Currently, there are a few specific efforts under way to forecast high winds and either alert drivers of high-profile vehicles or close wind-prone stretches of road.

Severe weather conditions, such as hurricanes, tornadoes, and flash floods, claim the lives of many drivers each year and cause numerous problems for traffic managers, law enforcement officials, and emergency managers (Figure 1-3). Hurricanes and flash floods can create a broad range of logistical challenges when large numbers of residents need to be evacuated from impacted areas in a short time. Such severe events also can make roads and bridges impassable and impair communication capabilities.

FIGURE 1-3 Hurricane Isabel begins to affect the U.S. mainland on September 17, 2003. The hurricane led to evacuations in North Carolina and Virginia and caused severe flooding as far north as eastern Pennsylvania. The federal government in Washington, D.C., was closed for two days largely because of concerns about the safety of commuters on the surface transportation system. SOURCE: NOAA.

ESTIMATED IMPACT AND COST

Adverse weather and the ensuing degradation of road conditions are associated with over 1.5 million vehicular accidents per year. These events result in approximately 800,000 injuries and 7,000 fatalities each year (see Table 1-1). Although weather may not be the only cause of these automobile accidents, inclement weather most likely is a contributing and exacerbating factor. The injuries, loss of life, and property damage from weather-related crashes cost an average of $42 billion annually (Lombardo, 2000). In some cases weather-related crashes can involve large numbers of vehicles and cause long stretches of roadways to be closed (Figure 1-4).

Any deviation from ideal driving conditions poses a potential threat to the maximum operability of the afflicted road. For example, during heavy rain or wet pavement conditions, drivers may decrease their speed by 25 percent (Figure 1-5), causing a drop in road capacity of 10 percent. In a snow event, they may reduce their velocity by nearly 40 percent, which can

TABLE 1-1 Total Deaths 1989–1999 Due to Traffic Accidents Associated with Adverse Weather

Vehicle Type	Total Deaths	Adverse Weather Present	Percentage of Total
Passenger cars	312,620	42,585	13.6
Light trucks	190,271	26,221	13.8
Heavy trucks	56,278	9,346	16.6
Motorcycles	28,537	969	3.4
Buses	3,617	622	17.2
Total	591,323	79,743	13.5

SOURCE: Lombardo (2000).

result in a 25 to 30 percent reduction in capacity. When visibility is less than 300 m, drivers may reduce their speed by 15 to 40 percent.

Drivers endure over 500 million hours of delay annually on the nation's highways and principal arterial roads due to fog, snow, and ice (OFCM, 2002b). This conservative estimate does not account for considerable delay due to rain and wet pavement. Because highways are the lynchpin between various modes of transportation, local delays can actually initiate impedi-

FIGURE 1-4 A pile-up of 100 vehicles during a storm on March 6, 2003, on Interstate 95 just outside Boston, Massachusetts. Ten miles of the interstate were closed while the vehicles were cleared from the road. SOURCE: Boston Globe.

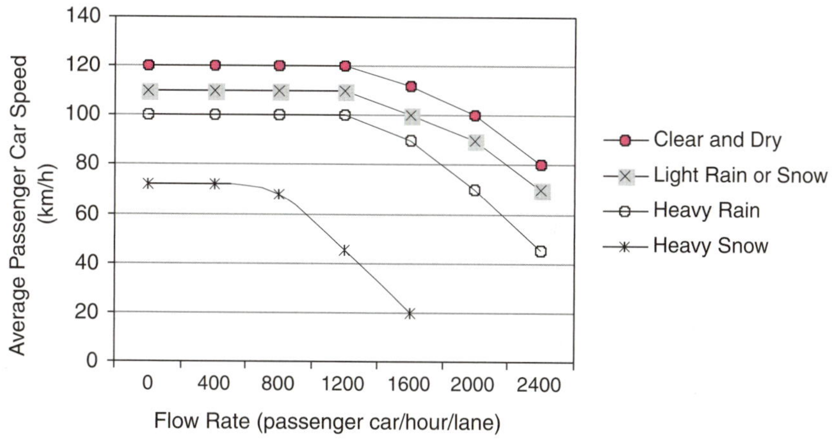

FIGURE 1-5 Speed flow curves for different weather conditions, assuming a free flow speed of 120 km per hour. SOURCE: Adapted from OFCM (2002b).

ments throughout the interconnected transportation networks. Such events can cause bottlenecks, interrupt delivery cycles, and create imbalances in supply and demand that lead to higher costs for businesses and ultimately for consumers. Given that the number of vehicle miles traveled is projected to increase substantially over the next 40 years (Figure 1-6) without a concurrent increase in highway capacity, adverse weather and weather-related road conditions will increase the strain on the roadway system.

Significant weather-related costs are also incurred by state and local agencies that maintain and operate the nation's roadways. For example, state and local agencies spend more than $2 billion annually on snow and ice control operations and over $5 billion annually for infrastructure repair due to ice and snow damage. The average expenditure for winter road maintenance in 1999 was over $22.7 million per state or approximately 12 percent of their maintenance budgets (FHWA, 2003). Although very few cost-benefit analyses of preventative techniques (e.g., anti-icing) or improved road weather information have been conducted, preliminary findings suggest the end justifies the means. For instance, Stowe (2001) reports a benefit/cost ratio of over two and a net benefit of over $1 million based on an analysis for a proposed automatic anti-icing system for a portion of a Washington State interstate. Another study that evaluated the use of anti-icing procedures by state highway agencies in nine states concluded that states

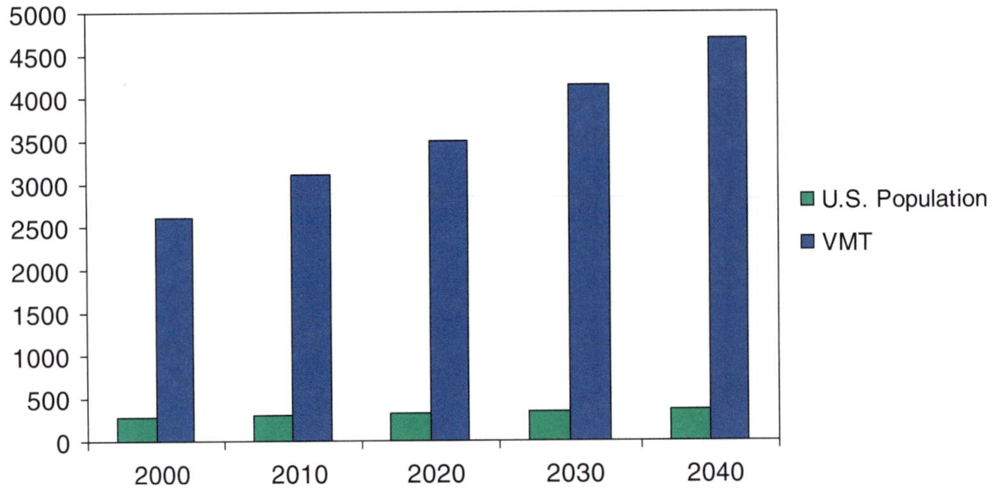

FIGURE 1-6 Projected U.S. population (in millions of people) and in vehicle miles traveled (in billions of miles). SOURCE: U.S. Census Bureau, 2000; TRB, 1994.

can reduce their snow and ice control budget by 10 to 20 percent (Boselly, 2001). Finally, in Kamloops, British Columbia, during the winter of 1996, the Insurance Corporation of British Columbia began exploring the use of anti-icing, de-icing, and pre-wetting practices during snowy or icy conditions as a supplement to the traditional use of salt and sand. As a result, winter maintenance costs were down nearly $15,000 (ICBC, 2000). Further use from 1997 through 2000 resulted in a 40 percent overall reduction in insurance claims related to snowy or icy conditions, a savings of over $4.3 million.

CHALLENGES IN ADDRESSING THE ROAD WEATHER PROBLEM

The fatalities, injuries, property damage, and economic costs associated with inclement weather present a significant road weather problem that is exacerbated by increasing traffic densities due to increasing population, vehicle ownership, and average annual miles traveled. This problem presents a number of challenges, both on a research level and when putting research advances into practice. The subject matter is highly interdisciplinary, spanning meteorology, technology, roadway construction and maintenance, traffic and emergency management, vehicle design and mechanical condition, and human behavioral factors that affect how drivers receive and process information. Developing new capabilities to deal with the road weather problem will require researchers who do not usually interact to collectively focus their attention on this problem. Moving the results of the research into an operational environment necessitates building relationships between researchers and a wide community of users. Many of these connections are only beginning to be made.

The roadway environment is very complex from a meteorological perspective, requiring knowledge of atmospheric conditions and the processes that influence them on scales much smaller than regularly considered in operational frameworks. Road weather conditions are influenced greatly by the immediate environment, including local terrain, built structures, and land cover. Hazardous conditions, such as fog or a frosty road surface, often develop in response to local conditions even when the weather forecast calls for clear skies; for example, pavement temperature will be colder at night in open areas compared to areas protected by nearby trees. These issues require accurate and timely meteorological data in order to effectively support the dynamic operation of the roadway.

Overlaid on the meteorological complexities is the need to consider the numerous factors associated with surface transportation. First, there are the challenges of the multifaceted roadway infrastructure; for instance, many different materials and methods are used to construct the roadway surface, all of which respond differently to various weather conditions. Second, there are several techniques for communicating weather information to drivers, who internalize and react to the information in ways that are not well understood. Third, the numerous vehicles that interact with the roadway infrastructure have different operating characteristics and weights, while drivers have differing diving skill. Finally, transportation personnel who build, maintain, and operate the roadway system employ a variety of approaches, which can alter the driving experience. Creating a more safe and efficient roadway environment requires consideration of all these factors under inclement weather conditions.

To address these challenges, the Federal Highway Administration requested the National Research Council to examine the research opportunities and required services needed to support improved weather-related information for the nation's roadways (see Appendix A for Statement of Task). The National Research Council formed the Committee on Weather Research for Surface Transportation: The Roadway Environment. The committee prepared this report after carefully investigating the current state of knowledge regarding road weather conditions. Following the discussion in this chapter of the affects weather has on the roadway environment, the committee's vision for the road weather system of the future is presented in Chapter 2. A review of the current activities and technologies relevant to the meteorological and transportation communities is provided in Chapter 3, including current observational and modeling capabilities of both weather and the roadway environment, forecasting capabilities, operational activities, and tools. Chapter 4 builds on the previous chapter by presenting the committee's recommendations for achieving the vision of the future operational road weather system based on identified gaps and new areas of needed research. In Chapter 5 the committee outlines a recommended coordinated national program to guide road weather research and transition scientific and technological advances into operations. The committee hopes that this report provides a useful framework to engage the transportation and weather communities and relevant stakeholders in the development of a strategic plan to guide road weather research.

2

The Road Weather System of the Future

High-quality weather information about the roadway environment, including both current observations and forecasts, communicated in a timely and effective manner can help drivers to make better, safer decisions regarding travel plans and to react properly when faced with potentially compromised conditions. Providing improved road weather information to those who construct, operate, and maintain the nation's roadways will enable them to operate the roads more efficiently and respond more quickly and appropriately to weather problems. Both drivers and transportation professionals will benefit from new technologies for vehicles, roadway infrastructure, and communications. In a more global sense, an exploration of road weather issues will help position the nation's road transportation system to respond appropriately to changes in demographics, technology, and the environment (see Chapter 6).

Recent and anticipated advances position the field of road weather for significant improvements in understanding and capability. For example, the meteorological community has made important advances in understanding smaller-scale meteorology. Great strides in observational capabilities have been made, making it possible to obtain a more comprehensive picture of current weather and roadway conditions. Numerical modeling has become more sophisticated and capable of resolving important phenomena on small space scales and short time scales. Transportation personnel have implemented dynamic message signs and have begun to take advantage of cellular technology to communicate with drivers while en route. Vehicle manufacturers have demonstrated the feasibility of a wide variety of onboard computing and telecommunications tools and have begun to move some of these to the market place. Accompanying these advances in meteorology and transportation are improvements in communications, computational capabilities, and geographic information systems, technologies that have applications to the road weather problem. It is clear that road weather is a problem that can be significantly mitigated by appropriate action, not a situation that simply has to be endured.

The idea of smart vehicles in constant communication with weather information providers and traffic control centers, commercial fleets constantly adjusting their routing to avoid anticipated storms, and road maintenance personnel being guided continuously by telemetered in-road sensors no longer needs to be limited to the realm of science fiction. Rather, over the course of the next 15 years a focused road weather research program could deliver this as a reality to the nation saving thousands of lives and billions of dollars. The committee sees the road weather system of 2020 as including a robust observation and communication infostructure,[1] models to support decisions, smart vehicles, enhanced roadway maintenance, and enhanced traffic and emergency management.

On her way to pick up her children from band practice Karen notes the dark clouds of a thunderstorm rapidly approaching. Meanwhile, computers at a private meteorological service are using data from the National Weather Service and the Kansas Department of Transportation to produce a highly specific "pathcast" for the storm, indicating which areas are expected to be affected. This pathcast takes advantage of a sophisticated four-dimensional data assimilation system that integrates local data from lightning sensors, radars, surface-observing stations, wind profilers, and stream gauges, to name but a few. Karen's in-vehicle communication system beeps three times and relays a message that the storm, which will be accompanied by lightning, gusty winds, and heavy rain with the possibility of flash flooding, will reach her location within 30 minutes. As the rain begins to fall heavily she worries about flooding of the nearby creek that already is high due to heavy rain during the past couple of days. Karen tells the hands-free communication system her destination and requests that it identify an alternate route to circumvent the nearby creek. The system integrates both weather and traffic information to send Karen on a safer route that will avoid flooded streets but remain uncongested.

[1] Infostructure is the network of data collection and dissemination necessary to support real-time management and operation of the roadway transportation system. It often is focused on congestion management, security management, emergency management, and weather response (Cambridge and Mitretek, 2003).

ROBUST OBSERVATION AND COMMUNICATION INFOSTRUCTURE

As illustrated by Karen's story the roadway system of the future will make a diverse collection of observations readily available to users when and where needed. Included in the observational network will be

- meteorological data from national and mesoscale in situ networks, satellite platforms, weather radars, and other remote-sensing instruments;
- an instrumented corridor for characterization of the road surface (for example using sensors embedded in the pavement) and the near road environment;
- vehicles that continuously measure conditions with which they come in contact, such as surface temperature, chemistry, and friction, air temperature, or visibility; and
- observations of land cover, stream flow, sea level, snow pack, and other environmental features that can affect the roadway.

These observations will be taken and then communicated to regional and national collection facilities, where they will be assimilated to provide continuously updated, four-dimensional fields of data. Periodically and on demand these fields of data will be used to initialize numerical prediction models, with the resulting forecasts stored in similar four-dimensional data fields. From these fields of observed and forecast data, interpolation techniques will be used to provide information at the location and time of interest to a roadway user. The interface to this vast quantity of observational and forecast data will be seamless. A robust "infostructure" will be in place that brings together all the desired data and disseminates the information to users, taking advantage of advances in communication sensor technology, information technology, global positioning system data, data management, computing, and geographical information systems, among others.

MODEL-BASED TOOLS TO SUPPORT DECISIONS

The road weather system of the future will employ sophisticated models to support enhanced decision making by the driving public, as in the case of Lou, and by those who build, maintain, and operate the nation's roads. The models will be "end-to-end," meaning they will integrate real-time observations of current weather, traffic, and road conditions; numerical weather predictions; models of traffic flow given these observed conditions; and rules of practice, to better serve a diverse customer base with decision

Lou is driving his semi-trailer tractor north on Interstate-35 across Iowa headed to Duluth, Minnesota. A major winter storm is heading east, with a foot of snow and strong winds forecast for much of northern Iowa and southern Minnesota. He consults his dispatcher using his "always-on" communication system to ask whether he should continue driving through the storm, stop and wait for the storm to pass, or head east and detour around the storm. The dispatcher must consider Lou's safety and the possibility that he'll be marooned for a few days if he continues, additional fuel and other vehicle costs, and any change in the time of delivery to the customer. Based on a continuously updated forecast that the storm is going to stay west of I-35, the dispatcher decides that Lou should continue on his original route. After a couple hours, however, the forecast now indicates the snow will spread farther east, causing patchy snow and ice accumulation on the road. Several vehicles about a mile ahead detect the icy conditions and report this through the highway's intelligent network. The main communication computer on Lou's truck picks up these reports, alerts Lou with a yellow light in the corner of his head-up display, and automatically reconfigures for better traction. Lou's dispatcher also monitors and relays to Lou information about the status of plowing and chemical treatments, automated spray of chemicals to anti-ice bridges, and traffic and visibility conditions, as well as the progress of the storm.

support, such as for snow and ice treatment or for pouring concrete. These models will be used by roadway maintenance officials, operational transportation managers, emergency managers, and those who build roads, and even incorporated into automated systems that provide alternate routes to drivers on the road. High-quality weather forecasts that provide information of the greatest accuracy possible at the necessary scale for the roadway environment are necessary to ensure the value of these modeling tools.

SMART VEHICLES

The cars and trucks of the future will be able to detect and respond to road weather conditions with ease. Not only will the vehicles be outfitted with instruments to measure road and atmospheric conditions but they will

Curt, the winter weather maintenance official responsible for western Montana, is awakened at 1:00 a.m. by his beeper; the forecast service has sent out an alert that snow expected to arrive in his region of responsibility is still on track and will begin shortly. He uses his personal information management system to access a suite of decision support tools, including highly localized weather forecasts, a sophisticated road temperature model, and an application that helps him choose treatment options. Curt is most concerned about two mountain passes that typically ice up first due to their higher elevation. He finds that three pavement sensors in a mountain pass confirm the previous evening's forecast that pavement temperatures would fall to below freezing just in that area. Curt quickly posts the weather radar display on his screen and the precipitation extrapolation shows that snow indeed is due at the mountain pass in a little over two hours. Based on this information, Curt quickly concludes that an anti-icing treatment would be necessary to ensure sufficient friction in the mountain pass in time for people to drive to work in the morning. A few verbal commands bring up his newly installed winter road maintenance decision aid system. It already has ingested all of the current weather and transportation information and confirms Curt's preventative maintenance strategy. It also recommends the precise amount of brine to apply for the current and forecast pavement temperatures and snow amount as well as the length of the roadway, based on road elevation, to treat on either side of the pass. Based on this information, Curt decides that an anti-icing treatment is warranted. He contacts the several employees he had placed on call the night before based on the weather forecasts he received, and by 2:30 a.m. they begin to apply liquid brine at the recommended rates to the road segments identified. The trucks are equipped with sensors for road temperature, air temperature and humidity, precipitation, and surface chemistry and friction along with a computer system that determines how much additional chemical treatment is necessary to address the conditions.

also have sophisticated positioning and communication capabilities. These "smart vehicles" could stay in constant communication with weather information providers and traffic control centers. Drivers will be informed immediately of suboptimum road conditions, such as whether a driver several minutes ahead encountered an icy road, or whether an accident occurred along the driver's route. Likewise, drivers will be able to call on route-finding tools to determine optimum routes, which will be of particularly great value for navigating roads affected by weather or congestion. The in-vehicle communication and route-planning tools will smoothly interface with World Wide Web tools available for pretrip planning, allowing planned routes to be transferred to the vehicle. For example, onboard navigation systems could be designed to show only the evacuation routes or to color code roads on the map to reflect up-to-the-minute driving conditions.

These smart vehicles will also use highly honed communication methods that take careful account of the human factors involved in receiving and processing information. They will be designed to provide the driver with an appropriate amount of information about, for example, quickly approaching hazardous conditions so that a good decision can be made. Innovative ways to communicate such information will be incorporated into the vehicle design, taking advantage of visual, auditory, and tactile methods. Some vehicles, as was the case for Lou's truck in the story above, will be able to automatically optimize its handling for specific road conditions.

ENHANCED ROADWAY MAINTENANCE

Efforts to remove snow and ice from the roadways in the past have combined regional scale coarse weather information, various rules of practice, lessons learned from experience, and a significant degree of guesswork. The road weather system of the future will be able to provide highly targeted weather information and decision support models that codify the best of past practices and use cost-benefit analyses to help maintenance officials make more informed decisions. Indeed, prototypes of this sort of decision support tool are already being developed.

Plowing equipment will also be much improved in the road weather system of the future. As was the case in the story about Curt, the plows will be outfitted with sensors that ascertain road surface temperature, friction, and even the amount of chemical already on the surface, along with software that automatically adjusts chemical treatment for the specific needs of small portions of roadway. In addition, equipment operators, who often struggle to see the road during heavy snow conditions, will be continuously guided by telemetered in-road sensors, magnetized lane tape on roads, or in-vehicle navigational systems.

The Sarno family attentively watches the television weather report in their south Miami home. A hurricane that developed north of the Lesser Antilles five days ago has intensified to a strong Category 1 with maximum sustained winds of 95 mph, and it is heading west-northwest toward Miami. Following the weather report an emergency management official again explains that evacuation is voluntary and that routes I-95 and I-4 have been designated as evacuation routes on which all traffic will be directed away from the storm. The Sarnos, who initially had decided to ride out the storm, change their mind. They quickly pack their hybrid SUV and prepare their home for the storm before beginning to drive to Orlando, where they plan to stay with relatives. After driving about an hour on I-95, an announcement comes over their in-vehicle routing system that the hurricane has accelerated and has taken a more northerly route than anticipated. Gale- and hurricane-force winds now are forecast to affect the coast and 20 miles inland—across the family's planned route—within 30 minutes and two hours, respectively. The Sarnos quickly query their in-vehicle communication system for an alternate route that has not been closed, is not overly congested, will take them farther west, and will avoid the windiest areas as well as those that are likely to flood. Computers at the regional transportation center compute a new route using information about all emergency- and construction-related road closures, real-time traffic observations, and geographic information system data. Following the new route, the Sarnos avoid the worst of the hurricane.

ENHANCED TRAFFIC AND EMERGENCY MANAGEMENT

Traffic managers in the year 2020 will have many powerful tools at their disposal for optimizing the capacity and efficiency of the roadway system. They will use sophisticated traffic simulation models that dynamically forecast how traffic would most likely respond to weather, construction, accidents, and other road closures. These models will take as input real-time traffic and weather data obtained from a much-enhanced observational infostructure. The traffic simulations will help identify ways to modify traffic flow and routing to respond to weather and other factors; for example, through changing signal timing and dynamic message signs that change the speed limit or advise drivers of detours or closures. The output also will provide guidance and recommendations directly to drivers before travel, when routes are being selected, and then in their vehicles once travel has begun.

Likewise, in 2020 responses to weather-related emergencies and evacuations due to weather hazards will take full advantage of advances in meteorology and traffic management through integrated decision support systems. This will be a significant improvement over the current "swivel-chair integration" approach, in which emergency managers consult separate information sources for weather, traffic, and emergency response practices. Further, these integrated decision support systems will be available to help determine optimum evacuation routes for individual drivers, just as the Sarno family was able to do en route.

3

Current Meteorological and Transportation Activities Relevant to Road Weather

Recent and anticipated research and technological advances position the field of road weather for significant steps forward in understanding and capability. Observational capabilities have made great strides, making it possible to obtain a progressively more comprehensive picture of current weather and roadway conditions in recent years. The transportation community is moving in the direction of a road network being operated as a "smart" adaptive system. Accompanying these and other advances in meteorology and transportation are improvements in communications, computational capabilities, and geographic information systems (GIS)—technologies that have clear applications to the road weather problem. In this chapter the committee highlights many current research and development activities that can be applied to road weather research, as well as other existing capabilities that have direct applications to road weather research but have not yet been fully exploited.

OBSERVING AND MODELING THE WEATHER

In Situ Meteorological Observations

Surface weather observations provide benchmark data about atmospheric and surface conditions to the scientific community and a broad spectrum of weather information users. The primary surface-weather-observing system in the United States is the Automated Surface Observing System (ASOS) that has been deployed over the past decade by the Federal Aviation Administration, Department of Defense, and National Oceanic and Atmospheric Administration (NOAA)/National Weather Service (NWS). There are nearly 1,000 ASOS sites across the United States; of those, 569 Federal Aviation Administration–sponsored and 313 NWS-sponsored sites are located at airports throughout the country (http://www1.faa.gov/asos/asosinfo.htm). These data are important in the verification of weather forecasts, in providing real-time weather information to the aviation community, and as input into data assimilation systems for numerical weather prediction. None specifically target the roadway environment. The ASOS is a fully automated system that provides an extensive suite of meteorological observations without human observers. It is sufficiently sophisticated to provide both routine hourly reports and special observations as warranted by changing conditions. The basic data given in each report include sky condition (clouds to 12,000 feet), visibility, present weather, surface pressure, temperature, dewpoint temperature, wind, and liquid precipitation amount. Although not routinely used, the ASOS has the capability to report as often as every five minutes. It was designed to support NWS warning and forecast operations and Federal Aviation Administration aviation weather needs; in addition, the system supports hydrological and climatological programs.

Despite its value to many users, the ASOS does not meet all users' requirements, largely because there are relatively few stations and because their observations are representative only of a small area near the site. As a result various user groups have developed and installed their own specialized surface-observing systems. Representativeness of surface observations is particularly important to the roadway environment, where minor differences in the physical environment (e.g., slope and exposure) lead to dramatically different effects. Although the ASOS provides useful data, it was never intended to be used to characterize the roadway environment; therefore, additional networks that target the roadway environment are needed.

Another very similar system is the Automated Weather Observing System (AWOS), which is a suite of sensors designed to collect and disseminate weather data primarily to assist the aviation community. The systems are classified as federal, which are owned and maintained by the Federal Aviation Administration, and nonfederal, which are owned and maintained by state, local, and private organizations. There are six different AWOS sensor arrays. The most basic array of sensors report wind speed (including gusts) and direction, temperature, dewpoint temperature, pressure, and density. The other five arrays build off this basic suite by reporting such additional parameters as visibility, sky condition, present weather, or lightning detection. Over 600 AWOS sites exist throughout the United States (http://www1.faa.gov/asos/awosinfo.htm). As with the ASOS, the AWOS was not deployed to observe the roadway environment, although its data are useful for synoptic weather observing and forecasting purposes.

The need for high-density observations is not unique to highway operations. Numerous mesoscale observing networks have been deployed around the United States, including the Oklahoma Mesonet (Horel et al., 2002b) and the Atmospheric Radiation Measurement Cloud and Radiation Testbed site (Stokes and Schwartz, 1994) in the central Plains. MesoWest, a heterogeneous collection of over 70 networks providing more than 2,800 observations, was assembled over the western United States in part to support the 2002 Winter Olympics in Salt Lake City, Utah (Horel et al., 2002a, b). These, as well as other networks operated by federal, state, or local governments or private entities, have been installed to serve special needs: most often meteorological research, agricultural operations, or air quality monitoring.

In addition to the surface-based observations, in situ weather data are routinely collected on commercial aircraft as part of the Aircraft Communication Addressing and Reporting System (ACARS) (Moninger et al., 2003). This program is the largest and longest-running data collection effort developed specifically for constantly moving platforms; it provides 80,000 reports a day with critical information about temperature, humidity, and wind in the atmosphere up to 15 km altitude (Figure 3-1). A similar system for

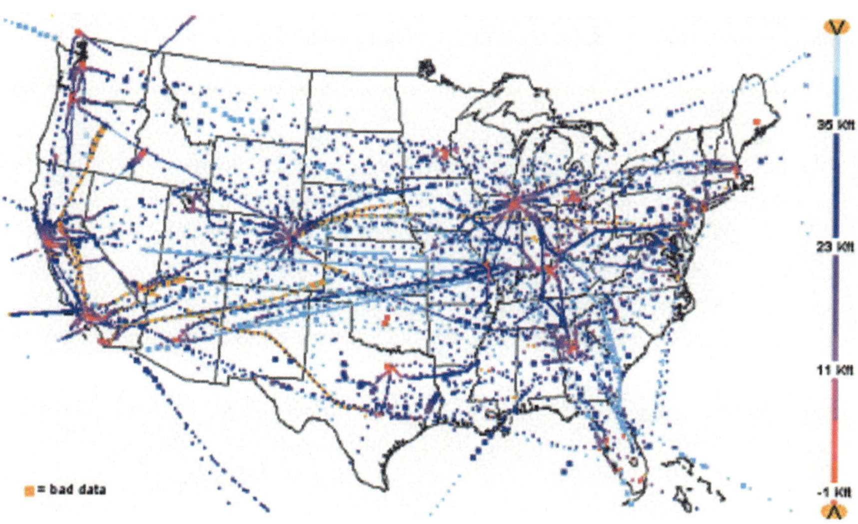

FIGURE 3-1 Locations of observations obtained from aircraft and collected in the Aircraft Communication Addressing and Reporting System (ACARS) for a 3.5-hour period in March 2000. The data are color-coded to reflect the height where they were obtained. SOURCE: NOAA.

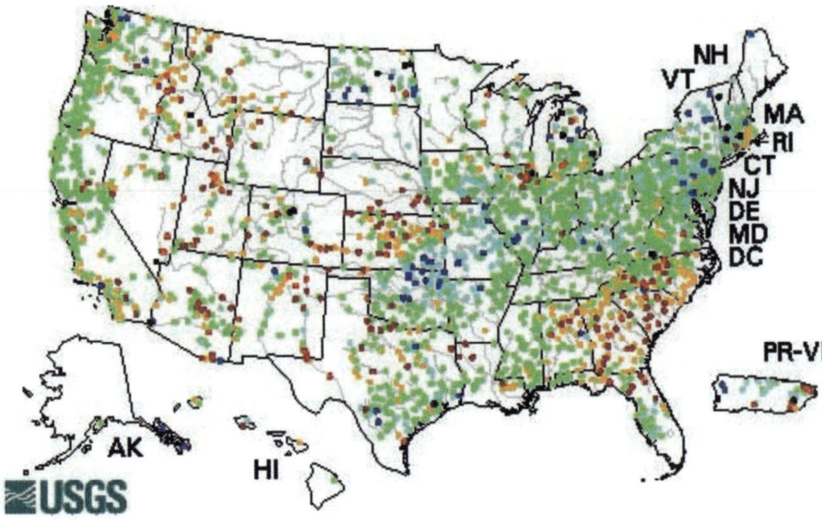

FIGURE 3-2 Daily streamflow conditions for January 21, 2004. Black indicates a new record high for the day, blue indicates flow greater than the 90th percentile, light blue indicates flow between the 75th and 89th percentile, green indicates flow between the 25th and 74th percentile, orange indicates flow between the 10th and 24th percentile, dark red indicates flow less than the 10th percentile, and bright red indicates a new record low for the day. SOURCE: U.S. Geological Survey.

observing the roadway environment using vehicle probes is described in "Observing and Modeling the Roadway Environment" later in this chapter.

In addition to in situ sensors that monitor meteorological conditions, there are networks that monitor responses to those conditions. Of greatest concern to the roadway environment is heavy precipitation, which can lead to flash flooding. The U.S. Geological Survey operates a nationwide network of gauges to measure streamflow, data that are available in near real-time (Figure 3-2). These data can be used—as in Louisiana's HydroWatch system (Wolshon and Levitan, 2002)—to gauge flood threat and if correlated with road elevation information, to determine when roads may become submerged.

Remote-Sensing Observations

Instruments that can observe atmospheric or land surface properties remotely offer ways to extend the spatial coverage of the observation network. These instruments can be "active," in which case they send out a pulse of electromagnetic energy and then use the reflected signal to determine characteristics of the atmosphere, or they can be "passive," in which case they measure radiation emitted naturally by the atmosphere.

Radar (Figure 3-3) is a widely used active remote-sensing technique that provides near real-time observations of the atmosphere under both clear-air and precipitating conditions. The deployment in the 1990s of Next Generation Weather Radar (NEXRAD) has provided the United States with a national remote-sensing network using Doppler radars (NRC, 2002). The radars provide nearly continuous monitoring of precipitating, severe weather complexes, and, when operating in clear-air mode, of nonmeteorological echoes (e.g., insects, dust), which can indicate wind speed and direction. NEXRAD incorporates sophisticated signal processing to sense the Doppler shift in the echoes returned from

FIGURE 3-3 A Doppler Radar. SOURCE: Bob Baron, Baron Services.

moving scatterers, thereby providing information on the wind field in the observed region. A new "volume scan" of a three-dimensional region around each radar is available every five to six minutes.

Efforts are currently under way to (1) integrate other radars producing a weather signal (particularly those operated by the Federal Aviation Administration) in order to enhance coverage and provide redundancy; (2) upgrade the radar to improve its capabilities for sensing precipitation; and (3) make higher-resolution radar data and derived products more generally available. Additionally, dual-polarization radars (i.e., radars that transmit and receive both horizontal and vertical polarizations) have the potential to aid weather observation and prediction by distinguishing between rain and hail and by identifying the precipitation type in winter storms. These radars could be installed in the national radar network within 5 to 10 years. Also in development are high-resolution national composites or mosaics, which will merge all available radar data into one database for the nation. Current regulations prevent NEXRADs from being operated at beam elevations below 0.5°, limiting the extent to which the radar can see very low elevations. Despite this institutional limitation, the NEXRAD is able to track precipitation, determine winds, and detect other phenomena such as blowing dust; thus it has the potential to be applied effectively to improving road weather information products (Mahoney and Meyers, 2003).

Profilers are vertically pointing Doppler radars that collect temperature, moisture, or wind data through the atmosphere. For example, 400-MHz-band wind profilers are able to detect winds directly above the profiler site at heights from about 500 m to 16 km, making them useful for weather forecasting. Additionally, there are 900-MHz-band radar wind profilers that can be combined with Doppler sodars to obtain boundary layer winds down to about 30 m. When wind profilers are coupled with radio acoustic sounding systems (RASS), temperature profiles down to approximately 1 km also can be obtained. A sequence of wind profilers for Conway, Missouri, is shown in Figure 3-4. Though profilers are a reliable, proven technology, there is not a dense network of these observations, and the lack of data at lower levels and the coarse vertical resolution limit their usefulness for near-surface applications (NRC, 2003c). Figure 3-5 shows the location of wind profiler sites in the contiguous United States. Recent advances in wind profiler technology have created the capability to monitor the altitude of the melting level in winter storms.

Observations from instruments on satellite platforms that actively and passively sense visible, infrared, and microwave radiation now routinely provide data that can be processed for information on the distribution of

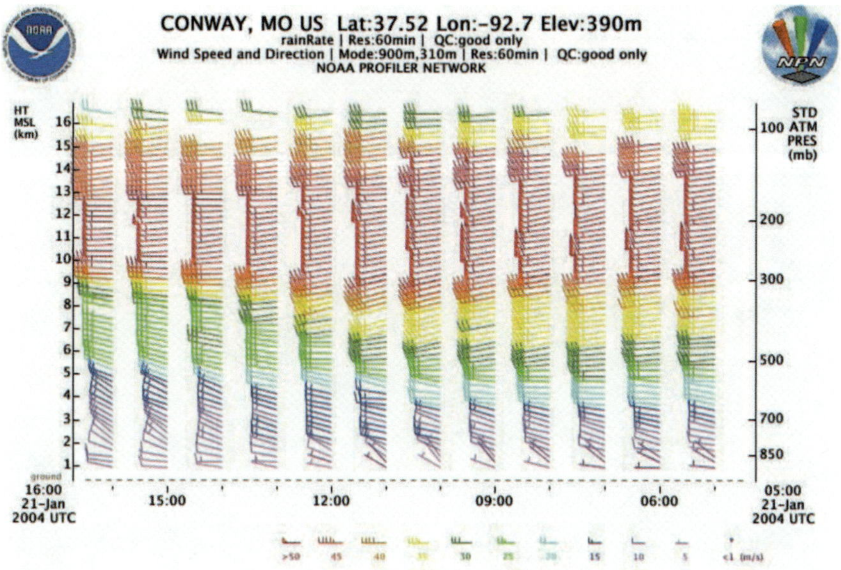

FIGURE 3-4 Hourly wind profiler observations over Conway, Missouri, on July 31, 2003.
SOURCE: NOAA.

atmospheric and surface parameters in both the horizontal (imagers) and vertical (sounders). One widely used satellite observation is water vapor imagery from NOAA's Geostationary Operational Environmental Satellite (GOES); such imagery often is animated to show the movement, development, and dissipation of large-scale weather systems. Satellite data provide excellent spatial coverage, filling in observations of weather conditions between surface-observing systems, and can be frequently updated. Civilian satellite sensor data are limited, however, in that they cannot resolve features the size of highways.

Several satellite-based sensing systems under development hold promise for applications to the roadway environment. Wind Index (WINDEX), an experimental GOES product, estimates the highest wind gusts that would occur if showers or thunderstorms were to develop. WINDEX is produced hourly from the sounder product and plotted on a satellite image. Originally developed for aviation operations, it could be used by surface transportation managers as an outlook to the potential occurrence of high wind and blowing dust associated with showers and thunderstorms, thus activating

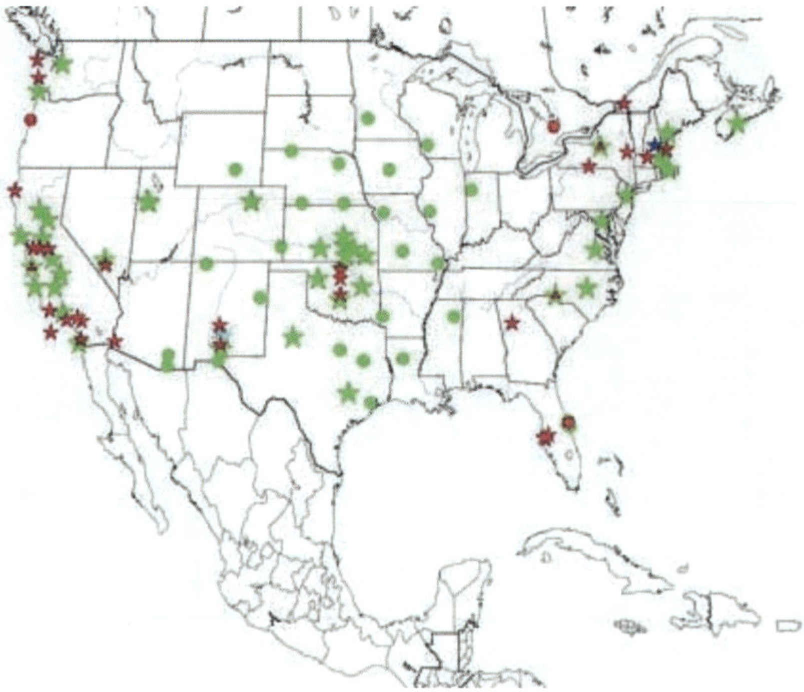

FIGURE 3-5 Distribution of wind profiler and radio acoustic sounding systems (RASS) sites in the contiguous United States. The star represents a wind profiler coupled with a RASS; the circle represents a wind profiler only; and the triangle represents a RASS only. The symbol colors represent data availability which vary daily. SOURCE: NOAA's Forecast Systems Laboratory.

cautions on dynamic message signs. On average, WINDEX estimates are within 5 knots of observed maximum surface wind gusts.

A fog and low cloud imaging capability, developed in support of aviation operations, analyzes different wavelengths of infrared radiation and using the differences, identifies low clouds and fog (Figure 3-6). Distinguishing between low clouds and fog is a challenging problem because there is a strong dependence on the underlying topography. Merging the satellite information with three-dimensional GIS data could lead to enhanced capabilities for distinguishing between low clouds and fog.

The Hydro-Estimator, one of the oldest quantitative derived satellite products, estimates precipitation down to the county level and, when combined with three-dimensional GIS data, could assist in anticipating proba b

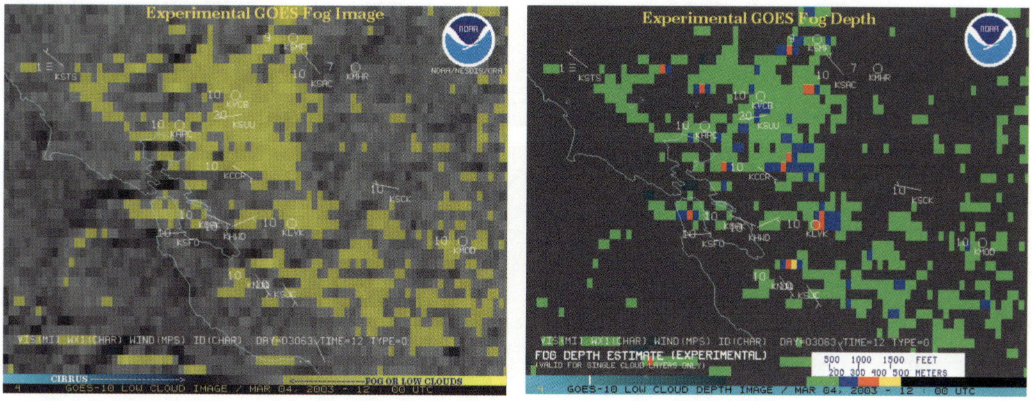

FIGURE 3-6 Fog location (a) and depth (b) over San Francisco Bay. SOURCE: NOAA.

lems in flash-flood-prone areas (Figure 3-7). These data are being incorporated in hydrological prediction models to forecast river and stream flows.

Monitoring the vast tropical oceans is a job that satellites do very well. GOES allows large regions of the tropical ocean to be continuously monitored throughout the life of a tropical cyclone, from genesis to dissipation. When in situ monitoring of tropical cyclones[1] by aircraft, ocean buoys, or ship reports is not available, minimum pressure and maximum winds can be estimated in a variety of ways using remotely sensed data. Most commonly, tropical cyclone intensity is estimated from cloud patterns and temperatures using visible and infrared satellite imagery. Other instruments, such as the Advanced Microwave Sounding Unit, the Special Sensor Microwave/Imager, and scatterometers, are being used to estimate intensity and wind structure. GOES data provide frequent updates of track and intensity changes as the tropical cyclone moves toward land, contributing to substantial improvements in the forecast of land-falling tropical storms and hurricanes. A new product, the Tropical Rainfall Potential, utilizes both GOES

[1]"Tropical cyclone" is the generic term for all tropical low-pressure systems, including tropical depressions, tropical storms, and hurricanes. A tropical cyclone is named when it reaches tropical storm intensity (maximum sustained winds of 39–73 miles per hour), and it becomes a hurricane when reaching maximum sustained winds of 74 miles per hour or greater.

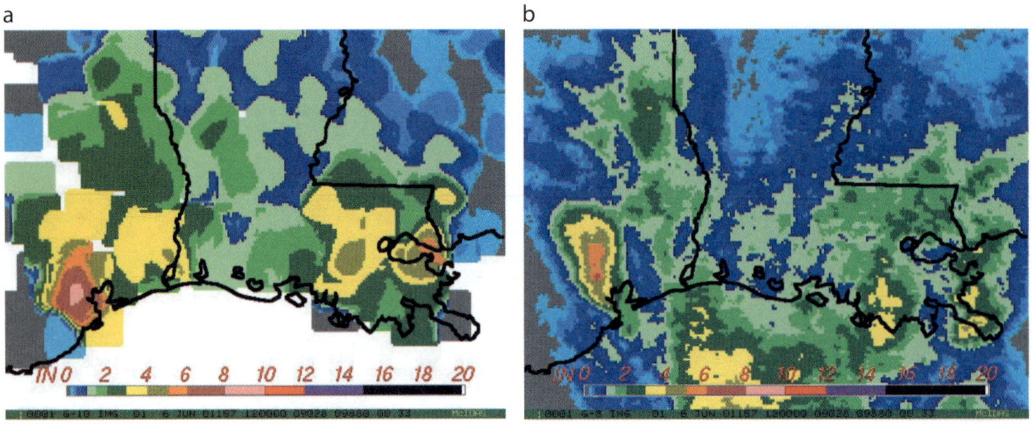

FIGURE 3-7 Satellite and radar precipitation estimates from tropical storm Allison: (a) in situ observations from rain gauges and (b) satellite observations from the Hydro-Estimator. Each panel shows total rainfall in inches for the 24 hour period ending on June 6, 2001 at 12:00 UTC. SOURCE: National Aeronautics and Space Administration.

and microwave data from polar satellites as well as the storm track forecast to estimate the potential rainfall of a system when it makes landfall. Ideally, data from satellites, radars, and rain gauges should be combined to provide the best estimate of rainfall.

Geostationary and polar satellites can also provide other variables of potential relevance to the roadway environment. GOES sensors can monitor land temperature changes under clear sky conditions. Other sensors on recent satellite series (e.g., the Polar Orbiting Environmental Satellite, Earth Observation Satellites, Defense Meteorological Satellite Program) can monitor a variety of surface properties daily to weekly. For instance, the Normalized Difference Vegetation Index is used to monitor vegetation greenness and health; these data can be correlated with blowing dust, which limits visibility, and they could be important for modeling the moisture flux of the roadway environment. A variation of the satellite vegetation product uses current weather information, particularly precipitation and temperature data, in a new experimental fire risk product that might provide additional insights for roadway managers regarding the spread and impact a wildfire might have on surface transportation and evacuation operations. GOES near-infrared data can be used to monitor the fire locations as well as the coverage and changes in smoke plumes (Figure 3-8).

a b

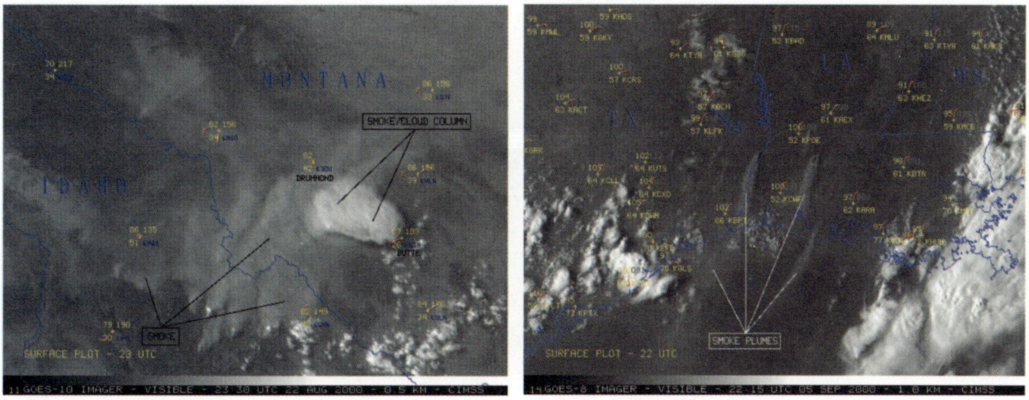

FIGURE 3-8 (a) Smoke from large fires in Montana and Idaho, August 22, 2000; (b) Narrow plumes from fires in Louisiana and Texas, September 5, 2000. SOURCE: National Aeronautics and Space Administration.

More than a hundred new sensors under development will be installed on satellite platforms within the next decade. Improved resolution, more frequent imaging, hyperspectral imaging, and new sounders are slated for launches by U.S. and global satellite partners. These sensors will provide new and improved observations of the atmosphere, including variables such as soil wetness, snow water equivalence, and smoke and aerosol detection over land. Examination of and development of applications from today's satellite-based observations will prepare the surface transportation community to maximize the use of these data now and in the future.

Modeling the Atmosphere

Numerical weather prediction is the foundation of modern weather forecasting. Today's state-of-the-science models are run multiple times per day and on horizontal grids with spacing as fine as just a few kilometers. The NWS currently is running the Eta model on a 12-km horizontal grid out to 3.5 days (84 hours) four times per day (http://www.nco.ncep.noaa.gov/pmb/nwprod/analysis/). Most forecast models are "full-physics" versions that provide explicit predictions of temperature, dewpoint temperature, wind, and precipitation. Forecasts for related sensible weather parameters, such as visibility and clouds, are generally derived by statistically post-processing the numerical model output, in some cases with direct human forecaster involvement.

Numerical weather prediction models have steadily improved since their inception (Figure 3-9). Accounting for this improvement are more accurate representations of subgrid-scale processes (e.g., clouds, precipitation, radiation), improved numerical methods including the ability to specify the initial conditions for the model using data assimilation, increased availability of observations, and increased power of supercomputers. When the accuracy of human-produced forecasts is assessed, it is noted that it tracks closely with the skill of the objective numerical weather prediction models, reflecting the strong dependence of human forecasters on the increased skill of numerical models (Figure 3-10). For the shorter-range forecasts extending out to three days, forecasters continue to add incrementally to the numerically generated forecasts using their knowledge and expertise.

The highest-resolution NWS models are currently run on horizontal grids on the order of 10 km, which limits resolvable features to those with horizontal dimensions greater than about 60 km. While this is sufficient to capture extratropical cyclones and the smoother characteristics of associ-

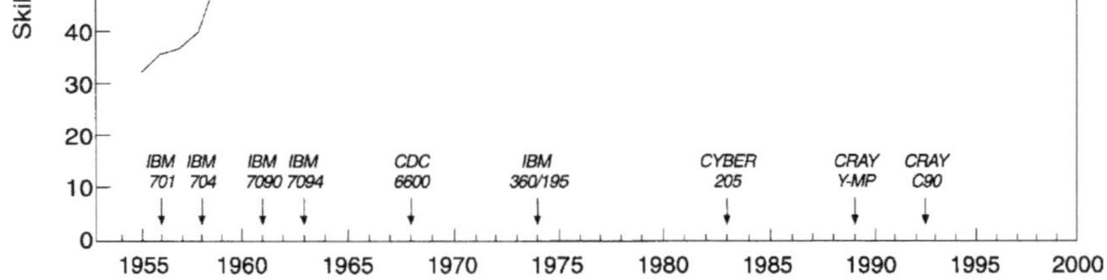

FIGURE 3-9 The skill score, an objective measure of the accuracy of numerical forecasts, of the National Centers for Environmental Prediction 500-hPa 36-hour forecast geopotential height has been tracked from the beginning of operational numerical weather prediction in 1955. A skill score of 100 is deemed perfect, where "perfect" means the average error is less than 20 percent. The complete history is presented in a review article by Kalnay et al. (1998). SOURCE: NRC (2000).

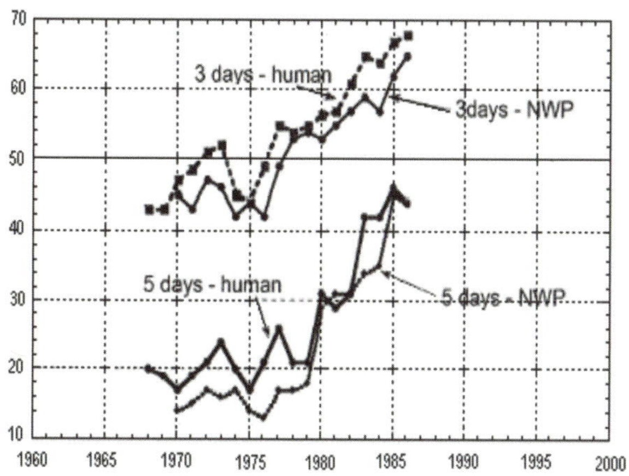

FIGURE 3-10 Relationship between the evolution of three- and five-day human and numerical forecasts. On average, the human forecasts are more skillful, but their improvement is strongly tied to improvements in numerical weather prediction forecasts. The score is the sea-level pressure anomaly correlation over North America, where the correlations are standardized to account for latitudinal dependence of the variability. SOURCE: Kalnay et al. (1998).

ated fronts, it is not sufficient to capture, for example, phenomena on the scale of individual thunderstorms, fine-scale precipitation bands, or localized wind events, all of which are critical to surface transportation. As a result additional applications or approaches are necessary to downscale to the domain of the highway. Such applications include additional numerical models at finer resolutions, physical models, statistical post-processing, or human-produced forecasts.

Regional mesoscale numerical weather prediction models, with grid spacings down to a few kilometers and that are typically nested within a larger operational model, are currently being run by university researchers, private sector forecasting businesses, and local NWS partnerships. These efforts are proving themselves to be useful for highway maintenance decision making. For example, a Pacific Northwest consortium of weather users has pooled partner resources to develop and support a real-time mesoscale modeling program (Mass et al., 2003). The Washington State Department of Transportation is such a partner and utilizes the 4-km grid forecasts for a number of transportation applications. To date, these efforts have been rather ad hoc in where they have developed, generally being driven by a combination of local interest, operational demand, and funding opportunities. Unfortunately, this approach leads to a lack of coverage, by any regional model, for some parts of the United States. Additional limitations include intermittent operations vulnerable to outages and uncertain funding.

The NWS is currently evaluating a new forecast presentation that involves a blending of numerical weather prediction forecasts, statistical and downscaling algorithms, and human forecaster decisions (Glahn and Ruth, 2003). The forecast is prepared using the Interactive Forecast Preparation System's Graphical Forecast Editor. This software allows forecasters to manipulate and edit grids, which can be initialized directly from a variety of operational numerical weather prediction models. A variety of graphical, tabular, and text products can then be derived from the gridded information. The forecast is composed on a fine horizontal grid (less than or equal to 5 km), has temporal resolutions on the order of 1 hour for the first few days, and includes a full suite of forecast parameters. Smaller domain grids prepared at local forecast offices are combined to form a grid with complete national coverage and 5-km grid-point spacing called the National Digital Forecast Database. In addition to direct delivery to a number of weather-information users via the Web and other media, the system is being designed by the NWS to encourage private sector businesses and partners to extract weather forecast information for desired regions or domains. These could easily include highway transects and domains.

As computing resources improve, it becomes increasingly easy to run very fine spatial resolution mesoscale models on smaller and smaller computing platforms. While increased resolution has many benefits, it can present a mismatch between what is known about the initial state of the atmosphere and the resolution of the prediction. A skillful forecast requires good observations of the initial state of the atmosphere and surface boundary conditions, but most current atmospheric observation networks do not provide adequate spatial resolution of the initial conditions. An exception is the NEXRAD radar network, which has a resolution of a few hundred meters. Both experimental and quasi-operational models initialized with radar data have shown great skill in predicting the onset of thunderstorms, for example, on time scales of six or more hours. Work is currently under way to develop a national mesoscale model for both research and operations. Termed the Weather Research and Forecasting (WRF) model, this modeling system will likely use radar data as a primary high-resolution input. Additional data sets that are likely to provide value in high-resolution modeling include the ACARS, profiler, and satellite data.

One way to deal with uncertainty in the initial conditions and derive useful forecast information is to use multiple, equally likely representations of the initial state. There are multiple techniques to obtain these initial state conditions, and each one is then used to initialize the numerical weather prediction model for multiple model realizations. This approach to determining a forecast, known as ensemble forecasting, provides useful information about the possible future evolution of the atmosphere in a probabilistic sense. Ensemble forecasting contrasts with the traditional method of taking the best available model and running it until it loses skill due to the growth of errors introduced in the initialization.

Ensemble forecasting approaches are successful in extending the period of forecast skill and also provide valuable probabilistic information on the likelihood of different conditions occurring. Figure 3-11 shows the probability of precipitation amounts exceeding certain thresholds. The precipita-

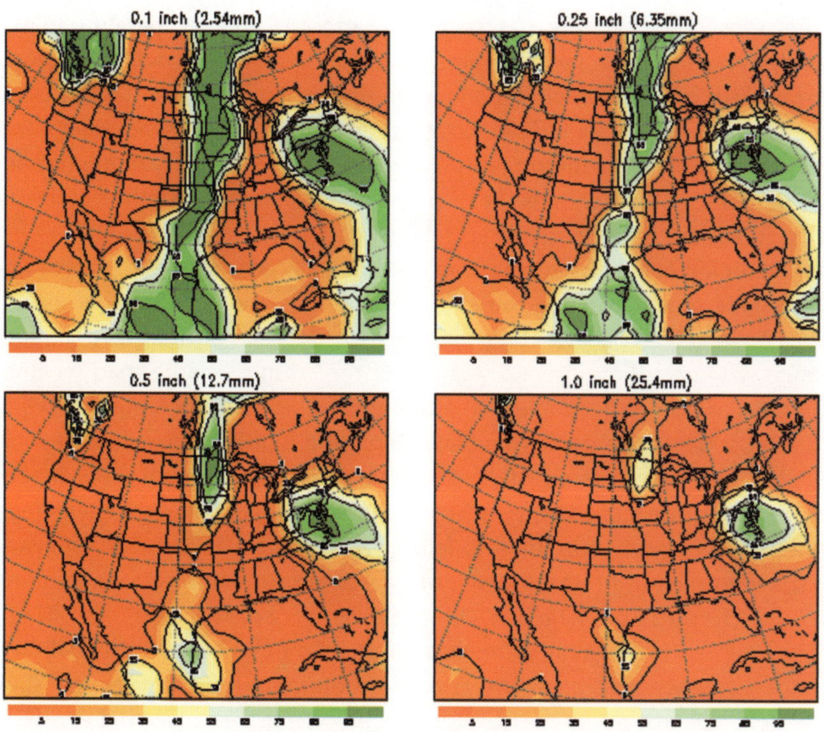

FIGURE 3-11 Ensemble-based probability of precipitation amounts exceeding thresholds of 0.1 inch, 0.25 inch, 0.5 inch, and 1 inch, September 18-19, 2003. SOURCE: NOAA

tion signature of hurricane Isabel can be seen along the mid-Atlantic coast. An American Meteorological Society (AMS) public statement "Enhancing Weather Information with Probability Forecasts" (http://www.ametsoc.org/AMS/policy/enhancingwxprob_final.html) was adopted by the AMS Council on January 13, 2002, and argues strongly for increased delivery of weather forecast information in probabilistic terms. This kind of information allows various user groups, including departments of transportation and transportation industries, to make cost-benefit decisions based on the likelihood of particular outcomes. These might include the threat of heavy snow, extreme cold, or critical thresholds of wind speed.

ROAD WEATHER FORECASTS

Operators and maintainers of roadways need specific, customized weather and road condition information in order to make the best operational planning decisions to provide a safe, efficient roadway system for users. Observed and forecast data needed most include temperature (pavement surface, air, dewpoint temperature), atmospheric conditions (cloud cover, visibility, and precipitation including probability, type, start and stop times, and amount including snow accumulation), and wind speed and direction. Ideally this information would be disseminated by common media (e.g., radio, cell phones) or displayed concisely in one integrated software interface that is easily interpretable.

Significant challenges remain in producing forecasts with spatial resolution on the scale of the roadway, but some efforts are already under way to provide targeted road weather forecasts for drivers and roadway maintenance workers. There are a few areas of the country where road weather forecasts are made available to the driving public. One such effort is being coordinated by the Washington State Department of Transportation in coordination with the Northwest Regional Modeling Consortium. The department joined the consortium, which has established meteorological real-time observing networks and modeling capabilities over the past 10 years, and added data from its own Environmental Sensor Stations (ESS) resulting in over 450 sites from which in situ meteorological data are gathered. These data, coupled with weather forecasts, are disseminated to provide real-time and forecast weather and road condition information to drivers and state maintenance crews with the objectives of helping drivers make better pretrip and en route decisions and providing roadway maintainers valuable information about the operability of the roadways. A pavement prediction model integrated with a land surface model also is available to provide current and forecast pavement temperatures to assist road crews with anti-icing decisions and operations. A single Web site (http://www.wsdot.wa.gov/Rweather/) provides these weather and model data as well as traffic incident and construction information, highway surveillance cameras, and highway advisory radio messages.

Another major effort supported by the Federal Highway Administration (FHWA) is the development of the Maintenance Decision Support System, a prototype tool for providing decision support to winter road maintenance managers. The prototype is being developed cooperatively by six national labs, which the FHWA selected according to their expertise and tools that could be applied to the problem. Based on the declared needs of the winter maintenance community, the Maintenance Decision Support System was designed to be an easily usable, single-platform decision support tool for displaying comprehensive information and providing recommended courses of action as well as the anticipated consequences of action or inaction. The system uses existing road and weather data sources and augments them as necessary to produce diagnostic and prognostic maps of road conditions with emphasis on the 1- to 48-hour timeframe. This timeframe is defined by the maintenance community's monitoring of incoming storms beginning approximately 48 hours before the event beginning, their activation of staff approximately 15 hours before, and their management (i.e., reassessment, coordination, and mitigation activities) approximately 1 to 3 hours before. During the winter of 2002–2003 the Maintenance Decision Support System prototype was deployed for field demonstration. Although only a few winter storms occurred, sufficient data were available to exercise the system. Field personnel responsible for winter maintenance operations felt that forecast accuracy must be improved before the system would be useful in improving the efficiency and effectiveness of winter maintenance control operations. Another, more comprehensive field demonstration is ongoing for the winter of 2003–2004.

Another project being supported by the FHWA with additional state departments of transportation, private, and international collaboration is FORETELL (http://www.foretell.com), an advanced weather prediction system that is being developed with the goals of reducing winter-related accidents and creating a viable road and weather information network. This Internet-accessible system acquires weather, agricultural, and road weather data to provide 10-km gridded 24-hour forecasts and nowcasts mapped to interstates and U.S. and state highways four times each day. FORETELL also uses advanced temperature models to calculate road surface temperature and predict future road conditions. After three seasons of test and evaluation, most elements of the program are working well, but forecast accuracy needs some improvement to be able to meet the high expectations of users.

In addition to the previously mentioned collaborations, the private sector also has been instrumental in leading the development of road weather forecast products. For example, the Advanced Transportation Weather Information System developed by Meridian Environmental Technology, Inc. was designed to provide decision support in planning and managing road construction and maintenance activities. The system includes multiple products, providing (1) site-specific weather and road forecast information for selected highways; (2) 36- to 48-hour area-specific forecasts; (3) current and forecast road restriction recommendations (e.g., to increase pavement life) based on freezing and thawing indexes combined with soil moisture data; and (4) a five-day site-specific, hour-by-hour weather forecast. Perhaps the most commonly used product is #SAFE, a phone system that served as a basis for the 511 system (described later), which drivers can call to learn about current and forecast road conditions along their corridor of travel. More information about these products can be found at http://www.meridian-enviro.com/products/trans.html.

Communicating Weather Information to Drivers

A variety of methods are available for communicating weather information to drivers. These include television, radio, and Web sites used to communicate weather information more generally, as well as mechanisms currently being developed to deliver targeted road weather information through Web sites, satellite radio, and the telephone, among others. Weather and traffic reports on television and radio are familiar to most. These reports typically provide current conditions and forecasts for metropolitan areas and substate regions, corresponding to their audience. Some television and radio stations have begun to take advantage of pavement sensor and weather data from ESS. These observations specific to the roadway might be displayed separately, rolled into an overall analysis of driving conditions, or used to make predictions, for example, when the pavement temperatures are well above freezing, that the road will not freeze for the next few hours even though snow is falling.

Most states have developed road weather Web sites that display real-time ESS (Box 3-1), ASOS, and other relevant data, including pavement temperature, air temperature, wind direction and speed, dewpoint temperature, precipitation, pavement condition, and subsurface temperatures (Figure 3-12). These Web sites may also show real-time cameras that monitor

BOX 3-1

Many states have implemented access to real-time road weather information through their state department of transportation Web pages. A few examples are listed below.

Alaska—http://www.dot.state.ak.us/iways/roadweather/rwisindex.html
Illinois—http://gis.dot.state.il.us/idotgis/rwis/
Kansas—http://www.ksdot.org/burcompser/generatedreports/weather.htm
Kentucky—http://www.kytc.state.ky.us/RWIS/index.htm
Maryland—http://www.chart.state.md.us//mapping/CHARTMap.asp?tab=Emergency
Montana—http://www.mdt.state.mt.us/travinfo/weather/rwis.html
Nevada—http://www.nevadadot.com/traveler/rwis/
Ohio —(standard site) http://www.buckeyetraffic.org/rwis/nosvg/
(enhanced site) http://www.buckeyetraffic.org/rwis/default.asp?w=1280&h=994
Washington—http://www.wsdot.wa.gov/Rweather/

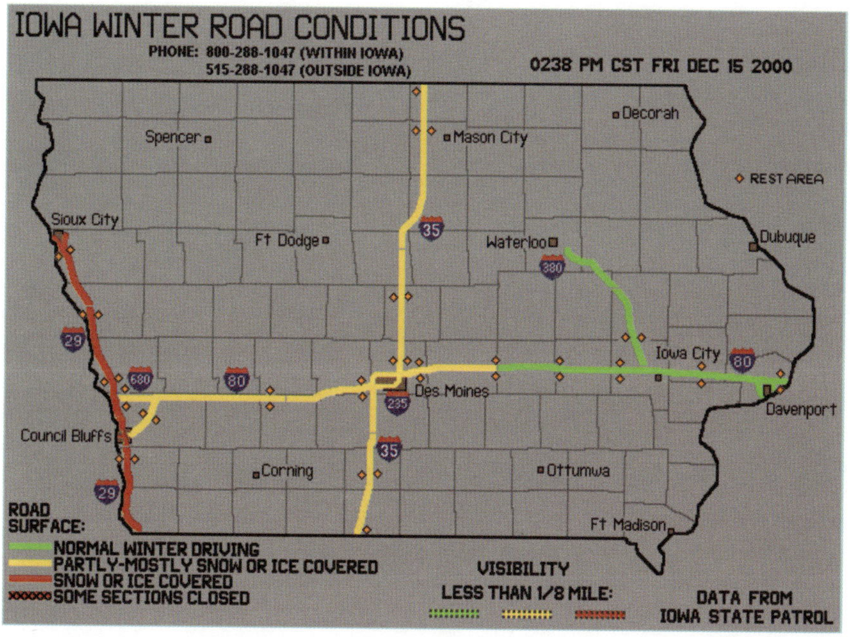

FIGURE 3-12 Example of a map from the Iowa Winter Road Conditions Web site (http://www.iowaroadconditions.org). SOURCE: Iowa State Patrol.

weather and traffic. Some of these Web sites are displayed in interstate highway rest areas.

In July 2000 the Federal Communications Commission designated 511 as a single national traveler information telephone number to be implemented by states and local jurisdictions. It is estimated that 511 will replace and consolidate nearly 300 travel information telephone numbers around the country. Great progress has been made in implementing this system, with most states supporting some level of activity (Figure 3-13). In snowbelt states where 511 is operational, the highest call volumes occur during the winter months when travelers use the service to receive road condition information. Ideally the system will be updated frequently so that near-real-time data will be reported. Many states are developing systems that allow users to interact with the system using voice recognition, which is a particularly important feature for cell phone users calling from the road.

Dynamic message signs are another important way to communicate to drivers (Figure 3-14). These signs have been in general practice for many

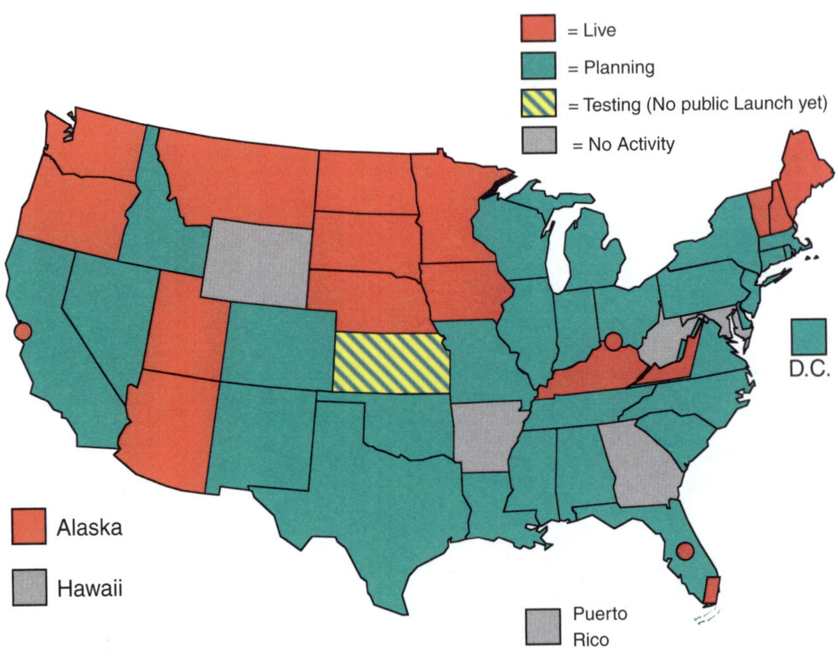

FIGURE 3-13 511 deployment status as of December 11, 2003. SOURCE: Resource 511.

FIGURE 3-14 A dynamic message sign. SOURCE: Delaware Department of Transportation.

years, and they use a number of permanent and portable technologies such as reflective flip-disk, light-emitting diode, and fiber optics. Drivers tend to pay close attention to these signs because they are displayed so close to traffic and have changing messages; however, it is not known how well drivers respond to such signs. Only a limited amount of information can be placed on the signs. Care must be taken to ensure that the optimum information is conveyed, at the right place, and with the end of a hazardous stretch of road signed as well. In some places, the signs advise motorists to dial 511 or tune into Highway Advisory Radio, a traffic information dissemination tool widely used by traffic managers, construction crews, and roadway maintenance personnel. Only limited use has been made of dynamic message signs to alert drivers of poor weather conditions, even though the signs are an attractive communication mechanism because drivers do not need to take any special information retrieval action. Although Highway Advisory Radio often has a limited broadcast range and poor sound quality, its appeal is its immediate and simple accessibility. Nearly every vehicle has access to AM and FM radio, so it is immediately available to the average driver who does not invest in new, specialized technologies.

Special Communication Needs and Tools for Motor Carriers

Roadway information services generally have been developed to most effectively meet the needs of the general motoring public. These services are sometimes limited to peak commuting times and travel routes, with information geared toward passenger car drivers in and around metropolitan areas. Trucking companies benefit from this information, but these commercial operators differ from the general motoring public in several ways, thus they require more specialized information to fully meet their operational and safety needs. Motor carriers operate 24 hours a day and 7 days a week; their trips often cover wide geographic areas comprising both metropolitan and rural areas. Routing and trip decisions usually are determined by a motor carrier's dispatch center, often in advance of trips and according to tight pickup and delivery schedules. Commercial drivers generally have less flexibility in certain aspects of their travel decisions (e.g., arrival time) than passenger vehicle drivers.

Motor carriers currently access roadway information through several channels. Dispatchers regularly poll information sources to determine whether trucks are heading into severe conditions. They then communicate with the drivers over citizens band (CB) radio or cell phone and route them around the problem. Drivers frequently use CBs to get information from other drivers, but do not rely on this information alone due to the large amount of erroneous information circulating the airwaves. Drivers also rely on the 511 system because the single number makes it easy to access information as they travel across state lines. A major concern for motor carriers and their drivers is the level of potential distraction of drivers while they are on the road; therefore, delivery of information into a truck cab should be as unobtrusive as possible. Drivers usually review the dispatch messages when they stop their vehicles. Dispatch centers prefer exception-based information; that is, information that falls outside normal operational expectations. Dispatchers are extremely busy and do not want to be distracted by information that does not directly require their attention.

An objective for improving communication in the trucking industry is the complete integration of roadway and weather information into a dispatcher's routing and dispatching software. This would allow all the information to be displayed in the dispatcher's daily work environment, thereby minimizing effort to retrieve and use it. Such an integrated system should include graphical maps showing color-coded trouble areas and comprehensive information in a one-stop format for the carrier's operating region.

More accurate information about traffic and weather conditions would have many benefits for the trucking industry. With accurate estimated arrival times, trucking companies could better coordinate pickup and delivery schedules to minimize wait time. When a truck is running behind schedule or is expected to face a weather delay, the dispatcher could call ahead and reschedule the pickup or delivery for more efficient loading or unloading of the truck. Customers could be informed when a shipment is expected to be delayed due to roadway conditions. With accurate real-time information and predictive capabilities, dispatchers could more accurately determine whether additional loads or stops were feasible, taking hours-of-service constraints and roadway conditions into consideration. Dispatchers and managers could use roadway information to verify whether drivers are unable to meet their schedules due to adverse road conditions.

OBSERVING AND MODELING THE ROADWAY ENVIRONMENT

Observing the Roadway Environment

In many areas multiple sensors have been installed to monitor critical roadway conditions. These sensors measure atmospheric, pavement, and water level conditions, and they are collectively referred to as Environmental Sensor Stations (ESS) (Figure 3-15). The complement of sensors can include a subsurface temperature probe, a surface sensor, a precipitation and visibility monitoring system, and atmospheric sensors that measure wind speed and direction, temperature, and relative humidity. Surface sensors, such as the one shown in Figure 3-16, can provide information on water coverage, snow, black ice, temperature, and even the amount of chemical present in a slurry on the pavement surface. There are over 2,000 ESS owned by state transportation agencies; over 1,400 are field components of Road Weather Information Systems (RWIS), which typically are used to support winter road maintenance activities. Field data from ESS are collected and processed by RWIS collection centers, where modeling and dissemination services are furnished to automated warning centers, traffic operations, emergency managers, and road maintenance facilities for decision support. Careful monitoring of the roadway is an important element of intelligent transportation systems (ITS), described in more detail later in this chapter.

Vehicle detection is an important observational component for modern traffic control systems. The most common technology used for vehicle detection utilizes inductive loop detectors installed in the roadway subsurface,

FIGURE 3-15 Environmental Sensor Station. SOURCE: Curt Pape, Minnesota Department of Transportation.

FIGURE 3-16 Pavement sensor. SOURCE: Jerry Waldman, Surface Systems, Inc.

which report the presence of a vehicle. Several newer technologies for vehicle detection allow for direct measurement of additional traffic parameters including traffic density, travel time, and vehicle turning movement. Most of these new technologies observe traffic remotely; for example, microwave Doppler radar can be used to observe vehicle speeds, passive infrared and acoustic detectors can determine vehicle presence or passage, and active laser radars can report vehicle passage, presence, and speed (Mimbela and Klein, 2000).

Video detection (Figure 3-17) of traffic and road conditions has become popular recently. With sophisticated image processing algorithms, informa-

FIGURE 3-17 Visual detector. SOURCE: Joseph Perrin, University of Utah.

tion about vehicle passage, presence, speed, length, and lane change movement can be obtained from video (Mimbela and Klein, 2000). Video data can also be presented on weather or traffic Web sites so that drivers and roadway maintainers can check road conditions. Additionally, video data have great potential for providing weather information, particularly visibility, presence of fog, and precipitation type. Algorithms to automate retrieval of weather information from video data are currently being developed (Hallowell, 2003).

Challenges abound in operating and maintaining road weather observation networks as well as in collecting and quality controlling the data. Currently there are no international or U.S. national standards for the basic ESS suite of sensors, nor for the siting, operating range, accuracy, survivability, calibration, and maintenance of these systems. Development of standards is being considered, but they are not yet mandated. However, communication standards for RWIS are being developed to exchange data between RWIS and other sensors and to communicate weather and road conditions to end users (FHWA, 2002). A more detailed description can be found in the section below on intelligent transportation systems.

In situ collection of road characteristics and environmental conditions by instruments on moving vehicles could evolve into a system similar to the ACARS system that collects weather observations on aircraft. Vehicle probes are available that can collect information on (1) roadway conditions, such as surface temperature, subsurface temperature, moisture, chemical concentration, and surface condition; and (2) roadway characteristics, such as the friction coefficient and road geometry. They also provide another source for in situ meteorological data, including air temperature and dewpoint temperature, wind speed and direction, precipitation, and visibility. With their inclusion in the National ITS Architecture, the exploitation of vehicle probes as opportunistic data sources is likely to increase. As exciting as this is, research is needed to determine how best to handle the millions of observations obtained, including who will organize and process the data, how it can usefully be incorporated into models and disseminated to users, and whether the most critical elements are being observed.

Modeling the Roadway Environment

Most current efforts to model the roadway weather environment use idealized energy balance models to predict pavement temperature. Energy balance models include weather information, heat and moisture exchange between the surface and the atmosphere, and the effects of precipitation, melting, freezing, and human activities (e.g., snow removal, application of salts to prevent freezing, and the traffic flow itself). The equation that describes the balance between sources and sinks of energy is

$$R_{net} = H + LE + G + Q_{melting} + Q_{precip+runoff} + Q_{anthro}$$

where R_{net} includes the downwelling radiation from the Sun and atmosphere, the reflected solar radiation, and the upwelling infrared radiation from the surface; H and LE represent heat loss from the road to the atmosphere by convection and evaporation, respectively; G represents warming or cooling of the road; and the source/sink terms Q account for the effects of melting, precipitation and runoff, and human modifications to the surface. Q_{anthro} includes heat originating from vehicles or the urban environment, or in the case of road construction, the heat released as the concrete or asphalt sets. Anthropogenic effects can also be included by modifying the other terms in the equation. Surface temperature can be determined by using an application of this equation. To estimate road slickness from frost or snow, one would combine a water-budget approach with temperature prediction that includes water phase changes and runoff.

Energy balance models have been used to predict road surface temperature since the 1980s when the U.K. Meteorological Office first used an energy balance model to predict the surface temperature of a dry road (Rayer, 1987). Such models have since become increasingly sophisticated and include the effects of water, ice, and traffic (e.g., Crevier and Delage, 2001; Jacobs and Raatz, 1996; Prusa et al., 2002; and Sass, 1992). Some of the complexities that can significantly affect road temperature are summarized in Table 3-1. An important limitation of energy balance models is their one-dimensionality; they are thus completely accurate only when surface conditions are horizontally homogeneous.

Energy balance approaches deal with energy exchange across the ground surface. Several approaches have been developed for roads in deep ravines, on steep hillsides, surrounded by high trees, and in other types of complex environments. Some researchers have approached this problem by solving the full three-dimensional energy balance for an area, as illustrated in Figure 3-18. These models tend to be computationally expensive and require substantial input data that are not readily available, and so they have not been widely applied.

Alternative approaches have been developed to investigate the complex roadway environment in models to predict surface temperature. Identifying scenarios and developing semi-empirical relationships or expert sys-

TABLE 3-1 Factors Affecting Pavement Temperature

Factor	Effect
Shading by road banks, vegetation, and bridges	Cools pavement during the day
Radiative heating in the infrared in steep terrain, from trees, etc.	Reduces radiative cooling at night relative to what is expected for well-exposed places
Thickness of pavement	Changes heat transfer into the ground and evaporation rates
Pavement composition	Changes heat transfer into the ground and evaporation rates
Pavement structure	Changes heat transfer into ground, evaporation rates, and runoff patterns
Structure and moisture of roadbed and ground beneath	Changes heat transfer into the ground
Traffic	Mixes air near the road; frictional production of heat from tires "grabbing" the road; radiative exchange between road surface and vehicle; heat and moisture fluxes from exhaust gases; air conditioning, etc.
Bridges	Vents beneath as well as above leads to more surface area to exchange heat with environment; cooler at night; variety of construction materials involved
Urban environment	Input of energy by city, known as the "heat island" effect; structures affect mixing and hence fluxes at surface
Clouds and fog	Affect radiative transfer (cooler during day, warmer during night)
Precipitation on the road	Phase changes affect heat exchange; creates additional interface
Road treatment strategies	Reduces snow cover, leads to melting, or prevents ice from forming
Road contaminants	Creates additional interface; affects heat exchange; affects precipitation on road

tems to predict hazardous road conditions have been useful. For example, Eriksson (2001) identified combinations of synoptic conditions and roadway sheltering by trees and high banks that lead to slippery roads, and Postgard (2001) developed a set of empirical relationships between changes in weather conditions and road surface temperature. Takle (1990) developed an expert system to predict roadway frost 20 hours in advance based on temperature, cloud cover, forecast early-morning dewpoint temperature, precipitation, and wind speed and direction. Although such approaches may not be universally applicable, they work well in particular regions.

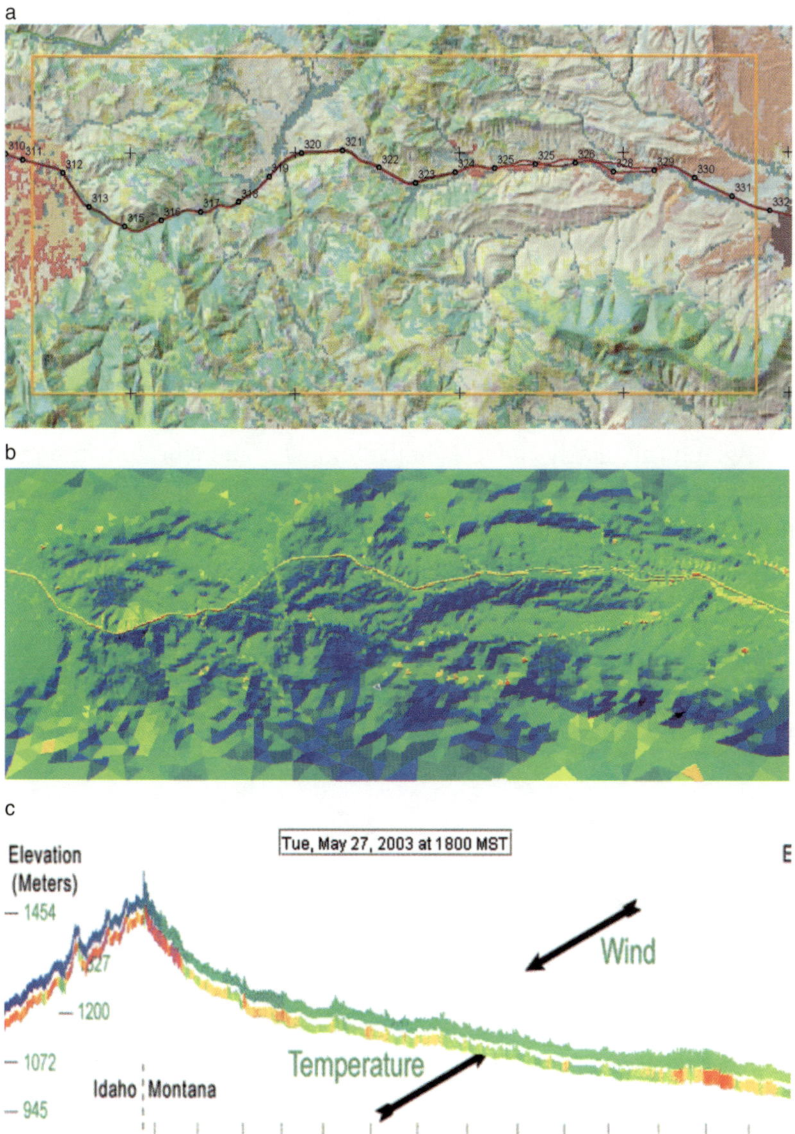

FIGURE 3-18 (a) Map of land type and elevation over the study area near where I-90 crosses from Idaho into Montana. These data are used as input to the three-dimensional energy-balance calculation. (b) Calculated land surface temperatures over the study area. (c) Road surface temperature and wind speed along the roadway as presented to the user community. SOURCE: Edward Adams, Montana State University.

ROADWAY OPERATIONS

Maintenance Management

To date, weather-related roadway maintenance activities have focused on responding to winter weather. The informational needs to support winter maintenance are fairly well defined (Figure 3-19). Snowplow operators, for example, need information that will allow them to determine the correct amount of chemical or abrasive to apply to the roadway. This decision depends on road surface conditions (e.g., temperature, chemical levels, and precipitation) during a 2- to 4-hour window, the typical time for a maintenance route. Inaccuracies in forecasts and the sensors that measure roadway conditions result in over- and misapplication of chemicals and abrasives, which adversely affects both budgets and the environment. Maintenance area supervisors, on the other hand, need to know the storm start time, the precipitation type and intensity, the storm ending time, road

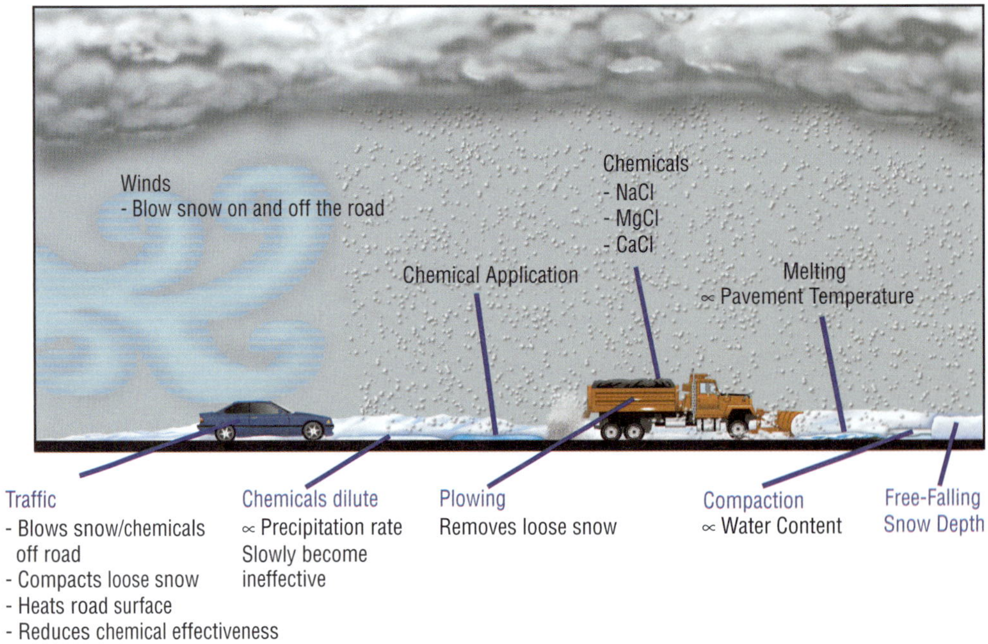

FIGURE 3-19 Winter storm treatment life cycle. SOURCE: Robert Hallowell, MIT Lincoln Laboratory.

surface temperatures throughout, and the weather conditions following the storm. Accurate storm forecasts are needed 18 hours ahead of the storm for the purpose of assigning shifts and mustering additional resources.

Current efforts build on the foundation laid in the early 1990s by the Strategic Highway Research Program (TRB, 2001a) and the Winter Maintenance International Scanning Tours conducted in 1994, 1998, and 2002. The highway operations section of the program contained a unit focused on winter operations research. The scanning tours created opportunities for exchanging snow and ice control technology and best practices with European countries and northern regions of Japan. These efforts have moved snow and ice control from being reactive, initiating operations only after the snow or ice begin to form, to proactive, beginning operations before the storm to prevent the bonding of ice or snow pack to the pavement. As implementation of the research results progressed, it became apparent that improvements in training, decision support, and equipment would be necessary to optimize the value of the research. This led to the pursuit of three ongoing activities:

1. innovative interactive training on Road Weather Information Systems and Anti-Icing led by the American Association of State Highway and Transportation Officials (AASHTO),
2. development and evaluation of the Maintenance Decision Support System led by the FHWA's Road Weather Management Program (described in the "Road Weather Forecasts" section earlier in this chapter), and
3. development and evaluation of the next-generation snow and ice control equipment led by a consortium of state departments of transportation.

With the introduction of new methods and technologies, winter maintenance managers and supervisors needed a better understanding of weather forecasting, snow and ice control chemicals, effective brine-making techniques, optimal application rates, dilution caused by weather and traffic, and pavement temperatures. Individual state agencies developed ad hoc training, but reductions in roadway maintenance staffs and the increased use of contractors resulted in a less experienced and smaller workforce. To meet these training needs AASHTO developed a computer-based, interactive, Road Weather Information System and Anti-Icing training program. Over 90 percent of the snowbelt states intend to use this training program, which will facilitate similar approaches to snow and ice control by their maintenance operators.

The correct snow and ice control equipment performing the right operations is needed to optimize winter maintenance operations. A consortium of snowbelt states led by the Iowa Department of Transportation has developed and field tested a new generation of snowplow truck. This truck is outfitted with off-the-shelf technology that uses real-time geolocated measurements of road temperature, surface chemistry, and surface friction, along with the ability to customize chemical application rates to predicted road conditions. This information is displayed for the plow operator and the supervisor so that both can monitor real-time operations. These automated features let the operator concentrate on driving the truck and allow the supervisor to monitor the entire fleet.

Traffic Management

Traffic management is the effort of roadway managers to optimize the efficiency and capacity of the roadway network. Over the past six decades the roadway system in the United States has expanded significantly, beginning with a U.S. highway system, followed by an interstate highway system and interconnecting arterials in metropolitan areas. The general public and commercial vehicle operators responded with ever-increasing traffic volume. New highway construction has now decreased markedly, yet the demand for roadway capacity continues to increase. The challenge to roadway operators is to manage the existing infrastructure with ever-increasing efficiency to produce a system that is safe with decreasing congestion and delays. Weather information is needed to improve roadway efficiency and capacity (Goodwin and Pisano, 2002).

Travel time, reliability, and delay are critical measures for system performance, both in real time and in archived applications. Traffic delay is being scrutinized in more detail to ascertain underlying causes. Weather is one of those causes as shown in Figure 3-20.

Traffic managers monitor roadway conditions and traffic incidents from traffic management centers. They use weather observations and forecasts from the commercial sector, including the media, to execute control strategies to manage traffic flow and advisory strategies to disseminate road weather information (Pisano et al., 2002). One control technique being employed in a few locations is the management of traffic signals when adverse weather is present or forecast (Perrin et al., 2001). When heavy rain, snow, slush, or icy conditions occur, operators access a traffic signal control system and manually implement longer-cycle signal-timing plans. Operators can further lengthen these cycles during peak commuting periods or worsening weather conditions. This technique reduces speeds and mini-

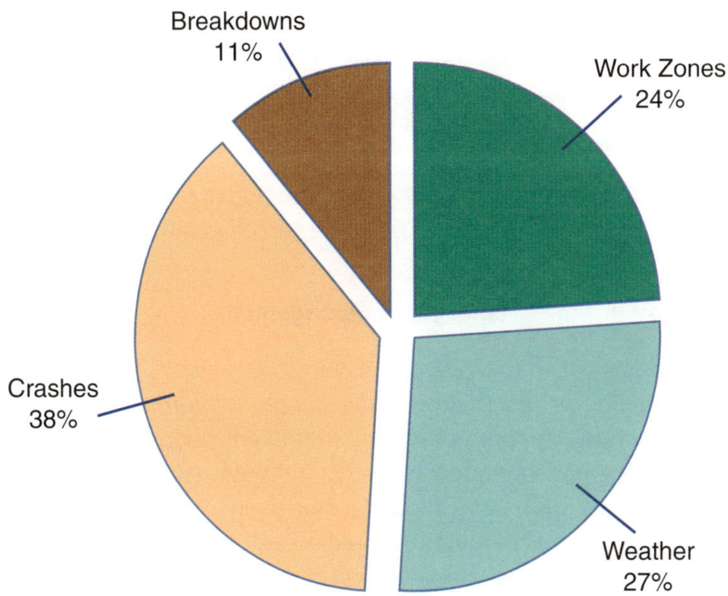

FIGURE 3-20 Causes of nonrecurring traffic delay. SOURCE: John Wolf, California Department of Transportation.

mizes the average number of stops per vehicle, thereby improving the safety and efficiency of the roadways. Once the road weather conditions improve, traffic signal operations are restored to normal.

Other control strategies used by traffic managers during inclement weather include speed management, land or vehicle restrictions, and guidance techniques (Pisano et al., 2002). Driver speed is managed based on the visibility, pavement, and traffic conditions by reducing speed limits to a safer velocity that is conveyed to drivers via dynamic message signs or variable speed limit signs. Traffic managers may also restrict access to road segments, lanes, or bridges or restrict the type of vehicles (e.g., tractor-trailers, vehicles without tire chains) allowed on portions of the roadway when visibility is low or lanes are obstructed due to snow or flooding. Finally, when visibility is poor, traffic managers may use pavement lights embedded in the road surface to delineate travel lanes, or they may use patrol vehicles with flashing lights to lead drivers safely through affected areas.

Although these traffic management tactics can greatly improve driver safety during inclement weather or poor road conditions, they are reactive rather than proactive. Better system performance relies on deployment of transportation management systems that account for weather (Nelson, 2002). Most authorities are quite familiar with the range of weather events that can affect their roadways, but only some organizations have mitigation strategies in place to exploit the use of advanced weather information.

Emergency Management

Major advances in intelligent transportation systems, weather research, vehicle technologies, electronics, and geographic information systems have created tremendous opportunities for improved emergency management practices for the transportation industry. Although emergency response to weather disasters such as tornadoes has been studied to some extent (Goodwin, 2003a), responding to tropical cyclones has received the most attention from the surface transportation research community because they call for massive evacuations relying primarily on highways. Efforts to address surface transportation implications of other weather-related emergencies are less developed and are not discussed in detail here.

The intensity of tropical cyclones and the damage they inflict when making landfall create considerable transportation-related problems (Figure 3-21). From the 1970s through the early 1990s hurricane activity was moderate in the Atlantic basin (Gray et al., 1993), and there was a concurrent explosion in population along the Gulf and East Coast regions of the United States (Pielke and Landsea, 1998). The rapidly growing coastal population and increased tropical cyclone activity necessitated thorough, efficient evacuation plans. The United States is among a small number of countries that rely on mass evacuations to protect their population during hurricanes. Evacuations used to be the responsibility of emergency management and law enforcement officials, but after the major traffic jams associated with Hurricane Georges in 1998 and Hurricane Floyd in 1999, the professional transportation community took a more active role in the planning, management, and operation of evacuations (Wolshon et al., 2001). These two hurricanes highlighted the need for better planning and coordination, increased evacuation route capacity, and better information exchange.

The most important aspects of effective hurricane evacuations are advance warning times and access to transportation; this is especially important because the majority of deaths associated with hurricanes result from inland flooding, and drivers are particularly susceptible. Track and intensity

FIGURE 3-21 Flooding during Hurricane Floyd, September, 1999. SOURCE: Louisiana State University Hurricane Center.

forecasts from the NOAA/NWS National Hurricane Center are used to monitor a hurricane and determine if evacuation is necessary. This decision usually is based on the hurricane's intensity (Table 3-2), size, track, and speed. A hurricane technically makes landfall when its eyewall—the most intense region of the storm—reaches land, but tropical storm- or hurricane-force winds can occur much earlier when strong outer rainbands reach land. Timing is of the essence when allowing for configuration of all traffic control elements on evacuation routes, the actual evacuation process, clearing of all routes, and removal of evacuation-coordination personnel once deteriorating conditions commence. According to Wolshon et al. (2001), medium-size cities need at least 12 hours to initiate and complete evacuation, but larger cities with limited evacuation routes, such as New Orleans, may require up to 72 hours. The preferred minimum evacuation-advance-notification times of several coastal states is given in Table 3-3. Although initiating evacuation earlier gives more time for people to leave, it also gives more time for the track of a hurricane to change; the average 24-hour track error during 1992–2001 was 93 miles (Franklin et al., 2003).

Several technologies and procedures have been explored and, to some extent, adopted to improve the efficiency of evacuation procedures. One issue that often has been overlooked is the interference of work zones and

TABLE 3-2 The Saffir-Simpson Hurricane Scale, which Rates a Hurricane Based on Its Maximum Sustained Winds

Category	Winds (mph)	Effects
1	74-95	No real damage to building structures. Damage primarily to unanchored mobile homes, shrubbery, and trees. Some coastal road flooding and minor pier damage.
2	96-110	Some roofing material, door, and window damage to buildings. Considerable damage to vegetation, mobile homes, and piers. Coastal and low-lying escape routes flood 2 to 4 hours before arrival of center. Small craft in unprotected anchorages break moorings.
3	111-130	Some structural damage to small residences and utility buildings with a minor amount of curtainwall failures. Mobile homes are destroyed. Flooding near the coast destroys smaller structures, with larger structures damaged by floating debris. Terrain continuously lower than 5 feet above sea level may be flooded inland 8 miles or more.
4	131-155	More extensive curtainwall failures with some complete roof structure failure on small residences. Major erosion of beaches. Major damage to lower floors of structures near the shore. Terrain continuously lower than 10 feet above sea level may be flooded, requiring massive evacuation of residential areas inland as far as 6 miles.
5	> 155	Complete roof failure on many residences and industrial buildings. Some complete building failures with small utility buildings blown over or away. Major damage to lower floors of all structures located less than 15 feet above sea level and within 500 yards of the shoreline. Massive evacuation of residential areas on low ground within 5 to 10 miles of the shoreline may be required.

SOURCE: http://www.aoml.noaa.gov/general/lib/laescae.html.

TABLE 3-3 Preferred Minimum Evacuation-Advance-Notification Times (in hours) for Several Southern and Eastern Coast States

	Hurricane Category				
State	1	2	3	4	5
Massachusetts	9	9	12	12	1
Rhode Island	12-24	12-24	12-24	12-24	12-24
Maryland	20	20	20	20	20
Virginia	12	18	24	27	27
South Carolina	24	24	32	32	32
Georgia	24-36	24-36	24-36	24-36	24-36
Mississippi	12	24	24	48	48
Louisiana	24	48	72	72	72

SOURCE: Wolshon et al., 2001.

the congestion caused by lane closures or detours. Most construction occurs during the summer, coinciding with the Atlantic tropical cyclone season, which extends from June 1 through November 30. To minimize this problem, some departments of transportation have included special provisions in construction contracts that require contractors to cease activities, clear their equipment, and reopen lanes in the event of an evacuation. Other states simply do not allow construction that reduces normal traffic capacity.

ITS information is being used increasingly during evacuations to collect and disseminate real-time data about traffic flow rates, road closures, weather conditions, and availability of alternate routes. Traffic cameras can be used to give visual confirmation of evacuation conditions. Highway advisory radio and dynamic messaging signs are being used to communicate important information to evacuees in a timely manner. Such information may include shelter locations, alternate evacuation routes, congestion and accident information, and services such as lodging and rest areas. The main limitation of such systems is that they most often exist in urban areas, whereas the majority of evacuation routes are in rural regions (Wolshon et al., 2001).

Since the traffic jams caused by hurricanes Georges and Floyd in 1998 and 1999, respectively, an evacuation procedure called "contraflow," which reverses one or more lanes of traffic flow to increase capacity, has been explored (Figure 3-22). There are four variants to this contraflow technique, assuming two inbound and two outbound lanes divided by a median (e.g., an interstate, a four-lane divided highway): (1) all lanes reversed; (2) one

FIGURE 3-22 Contraflow along I-26 near Charleston, South Carolina, during reentry after Hurricane Floyd in September 1999. SOURCE: South Carolina Department of Transportation.

lane reversed and one lane with inbound flow for emergency service; (3) one lane reversed and one lane for normal inbound flow; and (4) one lane reversed and the use of left shoulder of outbound lanes. The reversal of all lanes increases capacity by about 70 percent. Reversing just one lane and leaving the other for either emergency or regular inbound flow increases capacity by 30 percent, but it also increases the potential for accidents. Usage of the left shoulder for extra outbound traffic improves capacity by only 8 percent, and it has the potential for the greatest problems due to lack of pavement suitable for driving and the inconsistency of shoulder widths (Wolshon et al., 2001).

There currently are no standards or guidelines for designing contraflow operations. Typically a median is used for the crossover, but it also can take place at a freeway interchange. The decision about when to use contraflow, under what conditions, for how long, and how to communicate the information to the public usually falls on state governors, although they often task local law enforcement and state departments of transportation to manage the operations. The main criteria for deciding whether to use contraflow procedures are the characteristics of the hurricane (i.e., size, intensity, track, track speed), traffic volume, setup time, time of day, land use, and traffic conditions and patterns. By the time barricades are erected, traffic is cleared, and law enforcement is positioned, it can take 12 hours or more just to prepare a contraflow. Although contraflow operations are beneficial in providing increased evacuation capacity, the procedure can be inconvenient, confusing, unsafe, labor intensive, and difficult to enforce. Overall, the cost-versus-benefits remains unknown.

Several research efforts will be necessary for effective emergency management of the roadway environment. These include development of real-time detailed evacuation models that simulate actual roadway operations and combine various types of weather data with transportation and geographic data in one standardized display. Accomplishing this will require relationships between the meteorological and transportation researchers and practitioners to be further developed. In addition, these research efforts must be combined with tools for communicating reliable real-time information to the public, more complete data regarding the real-time operation of the transportation system, and improved mechanisms for sharing data and information. Institutional issues must also be addressed, as many different agencies and jurisdictions are usually involved in emergency management.

Roadway Design and Construction

Designers and builders of roadways must account for weather in their daily operations and long-term planning. For example, air temperature, humidity, wind, and precipitation play a fundamental role in the drying, hardening, and shaping of both concrete and asphalt. Ideally roads should be strong enough to carry loads, be durable and have low permeability to resist water and chemical penetration, be resistant to cracks and chemicals to prevent deterioration, and be aesthetically pleasing. However, for concrete, hot weather (75 to 100°F) can (1) accelerate setting, which inhibits a smooth finish; (2) increase the concrete temperature, which reduces the long-term strength of the pavement; and (3) increase the rate of hydration, which causes shrinkage and cracks. On the other hand, cold weather (< 40°F) can slow hydration, which retards hardening and strengthening (Smith, 2003). For asphalt the surface temperature before the mix is laid is critical, with higher surface temperatures (> 55°F) required for thinner slabs and cooler temperatures (> 35°F) allowed for thicker slabs (Spaid, 2004). Precipitation of any kind or amount can be very detrimental to concrete or asphalt construction. Liquid precipitation can create an imbalance in the

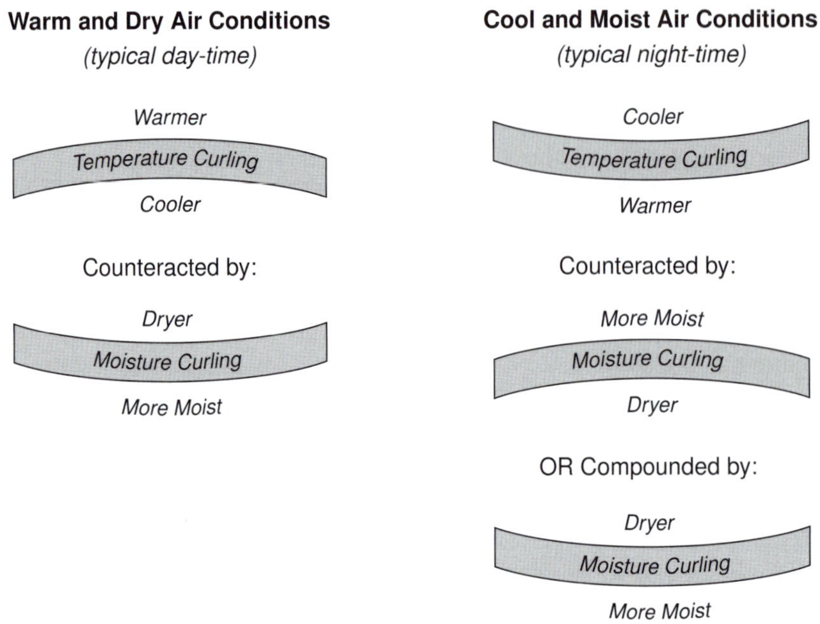

FIGURE 3-23 Schematic showing how temperature and moisture differentials above and below a concrete slab can cause it to curl up or down. SOURCE: Harold Smith, Center for Transportation Research and Education, Iowa State University.

moisture load, while hail can damage the surface, especially during the first 4 to 6 hours of placement. For asphalt, rain can wash away the tack coat placed on the surface before the mix is laid, leading to costly cleanup and replacement. Differences in temperature and moisture on either side of a concrete slab can cause it to curl (Figure 3-23).

The effect of weather on roadway construction goes beyond physical damage; it has an economic impact as well. Money can be lost even before paving begins if weather prohibits concrete or asphalt from being poured. For instance, concrete mixture can remain in the truck for only 30 minutes, so the mixture may have to be dumped if unexpected high winds or precipitation occurs. Inaccurately forecast or sudden storms also can result in lost wages for idle paving crews. Once the pavement has been placed, hail or rain can damage a slab to the extent that total removal ($8 to $10 per square yard) and reconstruction ($25 per square yard) are warranted. If only superficial damage is done, the concrete can be diamond grinded ($4 to $8 per square yard) to restore its surface (Smith, 2003). Even with notice of inclement weather there is some loss in production as work is suspended and the pavement is covered to minimize damage, then uncovered to resume paving. Despite such efforts, surface damage still may ensue. Ideally contractors want better than 85 percent accuracy with a minimum 0.5- to 2.0-hour notice for at least a 2 km^2 or smaller grid (Smith, 2003).

Most construction supervisors obtain weather information from the NWS, free Web-based services, and in some cases, specialized forecasts produced by private sector companies. However, the information often is not as timely or on as fine a scale as is needed by the roadway construction community. Decision support tools that meet the needs of the roadway industry are starting to be developed; for example, High Performance Paving is a prediction tool specific to the roadway construction community that uses temperature, wind, and humidity information to predict optimum paving windows and the impacts of adverse weather on roadway construction (see http://www.hiperpav.com). The weather information needs of the construction community were recognized by the inclusion of the Maintenance and Construction Operations user service in the National ITS Architecture in 2002.

INTELLIGENT TRANSPORTATION SYSTEMS

A major effort of the last decade by the surface transportation community has been to design and implement ITS. These systems take detection, computer, and communication technologies and apply them in an integrated fashion to increase the safety and efficiency of road transportation. For example, ITS is employed in Minneapolis and St. Paul, Minnesota, using a network of real-time freeway traffic detectors as part of a ramp metering system. The traffic flow information from the detectors is used as input to automated freeway capacity algorithms that regulate entry of additional vehicles onto the freeway system through modified traffic lights located at the entry ramps. Almost all state and local transportation authorities are using ITS to some extent today. ITS can be vehicle-based systems (e.g., adaptive cruise control, rear object detection) or infrastructure-based systems (e.g., vehicle surveillance, dynamic message signs). Typically it is the interaction between vehicles, the roadside, management centers, and travelers along with the synergy of data and information flows working in harmony that yield spectacular results.

Linking vehicles and infrastructure electronically is expected to advance rapidly in the coming years, as it is viewed as a very cost-effective means of improving safety and mobility. The application of sophisticated technologies in an integrated fashion across a number of transportation system components (i.e., vehicles, roadside, centers, and travelers) clearly requires a master plan. The transportation infrastructure (i.e., roadside and centers) in our highly decentralized transportation system must interact with vehicles of many types from many suppliers and travelers of all sorts needing to cover great distances across multiple jurisdictions. The U.S. federal government initiated extensive multistakeholder consultations in the mid-1990s leading to the development of the National ITS Architecture (National ITS Architecture Development Team, 2003). It is essentially a high-level blueprint for the application of information technologies to road transportation. First issued in 1996, the ITS Architecture continues to be refined. In 2000 the Architecture was adopted by Canada with some modifications. It is therefore well on its way to general use over most of North America, an enviable situation to promote freedom of movement and trade throughout the continent as well as collective security and global competitiveness. Indeed, the National ITS Architecture is a major body of work that will influence the evolution of road transportation for decades to come.

Enhanced use of ITS can mitigate some of the negative impacts of weather on road transportation by making it possible to provide the best weather and traffic information to users in a timely and appropriate manner (Andrey et al., 2003b; Cambridge and Mitretek, 2003). For example, in response to reduced surface friction during winter weather conditions, an advanced traffic management system could modify signal timing and ramp metering to adjust vehicle spacing accordingly, alert roadway maintenance crews to treatment needs, and inform drivers in their vehicles of potentially hazardous conditions. Version 4.0 of the National ITS Architecture, introduced in 2002, included the Maintenance and Construction Management Center and the Maintenance and Construction Vehicle to more formally capture road weather and the transportation components mandated to deal with it.

ITS has also been instrumental in the development of standards; for example, a group of communications standards referred to as the National Transportation Communications for ITS Protocol (NTCIP) includes standards for an ESS, NTCIP-ESS, which are open, industry-based standards that facilitate information exchange between RWIS and other ITS devices with a common communications interface. The standardization of terminology and graphical displays as well as the format and structure of messages are under development and will facilitate communication of road weather information to the public. The ESS sensors can be linked to an automatic spray system, perhaps developed by a different vendor, to dispense freeze-point depressants the instant that snow or icing conditions are detected on the road surface. Likewise, with sensors deployed according to agreed-upon standards in a coherent network over one or more jurisdictions, more sophisticated mesoscale modeling solutions could be pursued. Indeed, ITS provides a framework with which to extend road weather and other services in a fully integrated fashion across the nation, and eventually, throughout North America.

4

Opportunities to Enhance the Road Weather System

Many of the building blocks for creating the road weather system of the future are available today. Chapter 3 of this report summarizes the capabilities and technologies that are already being applied to road weather problems or that could easily be used. However, fully achieving the vision for a highly effective road weather system will require the research and operational communities to address a number of gaps in existing knowledge and infrastructure. These gaps at least partly reflect the lack of integration between ongoing surface transportation and meteorological activities. Thus, the committee recommends that a focused and coordinated national road weather research program be established to provide leadership in defining priorities, supporting the necessary research, and implementing operational infrastructure. With an improved overarching direction for the road weather community, there are substantial opportunities to produce improved products of interest to the user community through increased coordination of efforts (e.g., Nixon, 2001). At the same time there is a need to address fundamental scientific unknowns and make required observations that have been lacking.

As a framework for this program the committee identifies the following five key research and development requirements:

1. a robust, integrated observational network and data management system specifically designed to meet the need for enhanced road weather research and operational capabilities;
2. a coordinated research effort to increase understanding of road weather phenomena and options for increasing safety, mobility, and efficiency of the nation's roadways during all types of weather;
3. improved modeling capabilities and forecast tools designed to provide relevant, useful information to those who build, manage, maintain, and use the nation's roadways;

4. multiple mechanisms for communicating road weather information to varied users in ways that support better informed decision making; and

5. an infostructure that takes advantage of new technologies to effectively monitor and predict road conditions and then effectively convey road weather information to end users.

Opportunities for improving understanding, capabilities, and products in these areas of a new road weather research program are discussed in the following sections. The committee believes that advances in each of these areas will be necessary to attain significant improvements in road weather services and the benefits that are expected, as a result, in safety and efficiency on the nation's roads.

Coordinating this body of work will require a management structure that engages the multifaceted "road weather community." This community includes FHWA, NOAA, departments of transportation in each state, an important and vigorous private sector, the academic community, and professional societies (e.g., the American Meteorological Society) and other nongovernmental organizations (e.g., the American Association of State Highway Transportation Officials). One role of a national road weather research program would be to coordinate research, development, and implementation efforts undertaken by the various members of this road weather community. The recommendations in this chapter identify specific research needs that a new road weather research program should address, though in many cases the actual research, development, and implementation will fall to various members of the broader road weather community. Issues pertaining to management of a national road weather research program are addressed in Chapter 5.

OBSERVATIONS

A major objective of a new road weather research program should be to develop a robust, integrated observational and data management system specifically designed to meet the need for enhanced road weather research and operational capabilities. Many of the other objectives of the research program will depend on having implemented a road weather observation system, broadly defined; for example, improvements in understanding of road weather phenomena, modeling capabilities, and products and tools for end users require appropriate observations.

This road weather observation system should bring together a wide variety of existing data streams and new data as necessary to obtain a comprehensive picture of the roadway environment. Table 4-1 lists many of these observations. Several of the meteorological observations are currently used in an integrated fashion by weather forecasters and as input to numerical weather models. Other data, particularly those collected by the transportation community, are not yet integrated to the same extent or in a routine or standardized manner. Improved data management that enables efficient data sharing will be an important component of an improved road weather observation system.

Take Advantage of Existing Observations and Databases

As shown in Table 4-1, numerous observation networks are already available for assessing road weather; for example, many states have installed environmental sensor stations (ESS) along roadways to obtain targeted road weather observations. Data from ESS are essential to detecting and forecasting the effects of weather events on the roadway; however, they fall significantly short of the total amount of weather information needed to do this job well. The ESS data should be augmented with weather information from weather satellites, Next Generation Radar (NEXRAD) national radar network, surface sensors in other networks (e.g., Automated Surface Obsreving System (ASOS), aviation, agriculture, energy, water resources), vehicle probes, and weather data that can be derived from global positioning system satellites (e.g., see http://www.cosmic.ucar.edu).

Bringing all these data sources together allows a much more comprehensive picture of the roadway environment to be derived and used to provide information to drivers, roadway managers, and researchers. New technologies are rapidly emerging to assimilate data from disparate sources, but very little has been implemented to directly benefit road weather. Some private sector companies have pioneered the process of aggregating weather data from ESS and other weather observing networks to prepare targeted road weather forecasts.

In the case of winter road maintenance, data must become targeted specifically to the microclimate existing at or near the pavement surface. To accomplish this, sensor needs change significantly, and integration of other information, such as recent surface treatment history (e.g., plowing, sanding, and chemical application), become important factors. Treatment history collection and availability vary widely, but most of the larger state transportation agencies keep some type of records on time, equipment, and chemicals used in maintaining roads during snow and ice conditions. Methods for recording this information range from having operators manually record it to entering by means of computers when they input their duty time.

At the Minnesota Department of Transportation, for example, individual operators input this information along with their duty times using a Web-based application. Information on individual routes—including time expended, equipment and materials used—allows managers to analyze past actions, plan for future events, and determine whether current strategies are producing the desired results. Research in the areas of Automated Vehicle Location systems and Web-enabled cell phones is allowing road treatment and condition information to be reported more frequently. In the future it may become common practice to have in-vehicle equipment installed on snowplows to continuously measure and report this information.

Another untapped source of information are data on the characteristics of the surface and subsurface of the roadway that most state departments of transportation collect and retain during roadway planning and construction. This information is updated each time any construction or repairs are performed. Data on the road surface and subsurface composition are valuable inputs to pavement temperature models, in which the heat balance is affected by what is under the surface as well as what is above it. Research must be done to determine how changes in this profile affect the road surface and what would be the most efficient way to use this information to obtain more accurate pavement forecasts. Although many departments of transportation store this information in electronic format, they frequently identify it by mile posts, centerline distance, or stationing, which is significant only in its relationship to that particular road. In addition, different layers are identified by a numerical code that may be unique to that agency. If this information were to be used by an outside agency or vendor or shared among states, it would have to be converted to a geographic information system (GIS) format and layers identified with standard codes.

A final set of observations that could greatly enhance the road weather system is average and real-time traffic information. Data such as the average number of vehicles and average vehicle velocities, as well as real-time values of these parameters, are useful for strategic planning and reactive purposes. The speed and volume of vehicles influences the heat and moisture balance of roadways as well as their operability. These factors change diurnally (e.g., rush-hour traffic, early-morning hours), intra-annually (e.g., during prime vacation travel times), and during special events (e.g., concerts, sporting events). All departments of transportation have some type of statewide traffic-data-recording system composed of a remote-sensing system (i.e., pavement loops) and a data recorder that is used for monitoring and assessing traffic volume and speed characteristics (Wolshon et al., 2001). Surveillance cameras also monitor speed and volume, with direct visual confirmation. As with other previously mentioned observation net-

TABLE 4-1 Existing and Potential Observations That Should Be Applied to Road Weather Research and Operations

Observation	Description	Operated By	Road Weather Application	Limitations
Automated Surface Observing System	About 1,000 surface sites providing observations of sky and visibility, pressure, temperature, dewpoint temperature, wind, and precipitation amount	• Federal Aviation Administration • Department of Defense • National Oceanic and Atmospheric Administration/ National Weather Service	Provides weather observations used as input to numerical weather prediction models to ascertain real-time weather conditions on a mesoscale basis and aid in weather forecasting	• Observations are representative of small area directly adjacent to site • Mostly sited to meet aviation requirements • Lack of good winter precipitation observations
Automated Weather Observing System	Over 600 surface sites providing observations of wind data, temperature, dewpoint temperature, visibility, present weather, precipitation, cloud height, thunderstorms, and lightning	• Federal Aviation Administration • State, local, and private organizations	Provides weather observations used as input to numerical weather prediction models to ascertain real-time weather conditions on a mesoscale basis and aid in weather forecasting	• Observations are representative of small area directly adjacent to site • Mostly sited to meet aviation requirements • Lack of good winter precipitation observations
Various surface mesonets	Similar observations as ASOS operated on a regional basis with higher density of sites	• Oklahoma Mesonet • Atmospheric Radiation Measurement Cloud and Radiation Testbed • Many others	Provides weather observations used as input to numerical weather prediction models to ascertain real-time weather conditions on a mesoscale basis and aid in weather forecasting	• Observations are representative of small area directly adjacent to site • Sited to meet needs of specific mesonets • Data are not always easily accessible across mesonets • Lack of quality control and standards
Environmental Sensor Stations	A complement of atmospheric sensors adjacent to the roadway (air temperature and dew point, wind speed and direction, and optionally, pressure, precipitation occurrence, other precipitation	Individual states or municipalities	Provides most direct observations of roadway environment	• Observations are representative of small area directly adjacent to site • Individual networks are not linked • Insufficient standards for instruments and siting

TABLE 4-1 Continued

Observation	Description	Operated By	Road Weather Application	Limitations
	characteristics and visibility), pavement surface sensors (temperature, moisture, residual chemical composition, freeze point temperature), and sub-surface temperature probes			• Improvements in instrument technology needed
Aircraft Communication Addressing and Reporting System	Observations of temperature, humidity, and wind collected on commercial aircraft up to 15 km altitude	Aeronautical Radio, Inc.	Input to numerical weather prediction models and as aid in weather forecasting	Not directly relevant to the roadway environment
In situ observations from vehicles on the roadway	Observations of air and surface temperature, humidity, surface friction, and visibility in the roadway environment	Some prototypes being developed and tested, but not yet operational	• Information provided directly to drivers • Input to road weather forecasts • Input to numerical weather prediction models	• Challenges in communicating observations • Possibility for more observations than can be usefully applied
Weather radars, including NEXRAD	Nearly continuous monitoring of precipitation, including severe weather, and an ability to determine winds and detect blowing dust	• National Weather Service • Federal Aviation Administration • Private sector, including television stations • Military	Input into numerical weather prediction models (e.g., the Weather Research and Forecasting model)	• For NEXRAD, scans below 0.5° above horizon not permitted so areas distant from radar not well monitored • Network not very dense • Cannot discriminate precipitation phase (liquid or frozen) • Excessive ground clutter contamination from anomalous beam propagation • Conflicting user requirements

Continued

TABLE 4-1 Continued

Observation	Description	Operated By	Road Weather Application	Limitations
Wind and temperature profilers	Remotely sensed observations of temperature, moisture, and wind up to 16 km above the surface	• National Oceanic and Atmospheric Administration • Other state and federal agencies • Academia • Private sector	• Input into numerical weather prediction models • When combined with SODARS, the 900-MHz band profilers can be used for microscale modeling of the roadway • Improve weather forecasts	• Representative of a small area • Network not very dense
Geostationary Operational Environmental Satellite, Polar Orbiting Environmental Satellite, and other satellites	Observations at multiple wavelengths (e.g., visible, infrared, microwave) used to derive, for example, integrated water vapor and cloud-track winds, tropical cyclogenesis	• National Oceanic and Atmospheric Administration • National Aeronautics and Space Administration • Military	Improve weather forecasts	Cannot resolve features on the roadway
Traffic observations	Traffic detectors and cameras	Individual states or municipalities	• Managing the traffic system during inclement weather • Input to energy balance models • Study of weather impacts on traffic	• No standard format or reporting procedure • Expensive to store
Recent treatment history	Data record indicating the chemical application, plowing, or sanding of a given stretch of roadway	Individual states or municipalities	Input for forecasting road surface conditions and providing decision support for further treatments	No standard format or reporting procedure
Roadway surface and subsurface	Surface and subsurface sensors used to measure temperature	Individual states or municipalities	Input for energy balance models that forecast road surface temperature	Network not very dense

TABLE 4-1 Continued

Observation	Description	Operated By	Road Weather Application	Limitations
Streamflow	Measurement of the movement of surface runoff traveling in a stream	• U.S. Geological Survey • National Oceanic and Atmospheric Administration • Others	Useful for forecasting floods	Network not very dense
Landsat	Land surface cover/land type	National Aeronautics and Space Administration	Input for energy balance models that forecast road surface temperature	Maps and data require some user interpretation
Global Positioning System	Satellite-based system to show position on Earth's surface; can also be used to derive total water vapor column	Department of Defense operates the satellites; receivers are widely available	• Useful for providing targeted geolocated information • Derived water vapor useful for improving weather forecasts	Selective availability leads to degraded accuracy in position readings

works, if traffic data are to be used by other agencies or vendors, standards for displaying the information must be established.

Recommendation: Take advantage of existing observation networks and databases.

The committee recommends that the road weather research program take full advantage of established environmental monitoring networks, including those from in situ and remote-sensing platforms and existing databases on soil type and land use characteristics. As a first step, efforts are needed to integrate and use information from environmental sensor stations and meteorological mesonets with particular attention to seamlessness across state boundaries. A nationwide repository in which weather observations relevant to the roadways are collected and shared by all interested parties would be of great use to the road weather community.

The committee recommends that the road weather research program support the design and establishment of a portal that allows access to (1) weather data relevant to the roadway; (2) state databases of relevant road characteristics, including pavement or concrete composition, road surface depth, and underlying soil type; (3) real-time road surface condition information; and (4) traffic information. Access to these data currently is constrained by the limited data availability and disparity of data formats used by states and municipalities. Having access to these data in standard formats would allow private sector companies to develop forecasting tools that could be easily and reliably applied in different states.

Expand the Network

Exploiting weather information to increase the efficiency of the roadway system requires weather, road condition, traffic, and other observations that are representative of the roadway. Additional observations are required at locations especially susceptible to adverse weather conditions, such as valleys where fog often forms, frost hollows, or roadways subject to high winds. Over the last 20 years most states have implemented ESS networks to collect valuable weather data in targeted locations. At the same time, various meteorological observing networks have been established to meet general weather forecasting needs, as well as those of more specialized sectors, such as aviation, forestry, or agriculture. These observational efforts must be expanded and customized to fully meet surface transportation needs by collecting tailored information along the roadways, particularly at vulnerable locations susceptible to adverse weather conditions; for example, the National Weather Service (NWS) should consider road weather needs in planning new observation sites.

Although Road Weather Information Systems (RWIS) technology has been used for many years, research is needed to make optimal use of it; for example, only a few states have attempted to determine optimum sensor density. RWIS companies have recommendations for spacing of the sensors, but in most cases they are based on regional or national average estimates rather than site-specific analyses. Given funding limitations, roadway managers need refined methods to prioritize sensor locations in the network. A science-based method allows the determination of which sensors give the operator the largest value for the investment costs. Roadway managers also need methods of determining where to locate sensors, such as the optimal height above the road surface, distance from the roadside, or placement in the road surface; for example, in some cases it might be more advantageous to locate a sensor far from the roadway to detect weather conditions that influence that segment of the system.

Tapping into the data from vehicle probes would vastly increase the number of observations available for the roadway, much the way observations from aircraft have been exploited. Although the technology is already available, it is not used routinely. Commercial fleets represent an excellent opportunity for experimentation and evaluation of these data because they provide a representative sample from a manageable amount of data with tighter institutional control.

Observational technology ranging from soil and pavement sensors to universal vehicular highway reporting is possible, providing exciting opportunities for expanding knowledge of the roadway environment; however, determination of what is to be measured, and where and when, are challenges. Research is needed to explore the quantity and accuracy of data needed for mesoscale forecast models, decision systems, and highway maintenance personnel. Who will do the research and where it will be conducted also must be explored.

Recommendation: Improve the existing observation system.

To attain an improved road weather observation network the road weather research program should support efforts to establish standards for observing procedures and station siting to ensure that (1) the data are representative of the roadway environment and (2) new road weather stations and traffic monitors are added to optimally address both modeling and special observing needs. The National Weather Service, states, and other entities that perform weather observations should use these standards as they establish new sites to fill gaps in the current observational network.

In addition to improving fixed road-weather-observation sites, expanded use should be made of mobile observing platforms. Such platforms could consist of a fleet of state- or city-owned vehicles, commercial trucks and buses, and even private vehicles. Vehicle observations of roadway temperature and surface traction could be of great use in providing a more complete and accurate picture of roadway conditions. The road weather community should collaborate with the automobile industry in the design and deployment of these mobile observing platforms and ensure that these data remain open to access.

Instrument Technology

A final way to ensure robust observational capabilities is to make certain that the instrumentation provides high-quality, accurate data. Careful testing and calibration of sensors is an important step in improving instrumentation. Although testing and calibration standards have existed for atmospheric sensors for some time, current sensors, sensor standards, and instrument placement are tailored to meteorological needs and, in many cases, do not meet the specialized needs of maintenance, traffic, and emergency managers. For instance, precipitation sensors must be capable of accurately identifying precipitation type and amount as well as the liquid equivalent through all temperature ranges.

There currently are no standards for surface and subsurface environmental sensors. More concerted research is needed to test and compare the pavement sensors. To deliver the accuracy and response time required by sensor users, close attention must be paid to sensor design, installation, and calibration. This is needed to ensure that the thermal properties of the surrounding pavement are accurately represented by the sensor.

The majority of research and development of environmental sensors takes place in the private sector, which has led to new and improved sensors being offered by multiple companies; however, most of these sensors have not been rigorously tested and compared, meaning that it can take many years to design and get funding for an experiment that tests the sensors in a real-world setting, and then run the experiment long enough to get sufficiently representative weather conditions. By the time a research group has undertaken all of these steps it is likely that the sensors being tested will be outdated. An additional problem for testing pavement sensors is the difficulty of creating a controlled experiment as roadway maintenance engineers prefer to treat the roadway when called for. As a result, insufficient information is available to consumers when choosing which sensor to purchase. To date, environmental sensor users have relied almost exclusively on accuracy, installation, and calibration information provided by product manufacturers. While this information can be useful (e.g., Partanen et al., 2003), it is often produced in controlled laboratory or test-bed conditions, and therefore can be far removed from real-life conditions that affect sensor performance. There are currently two projects under way to test environmental sensors in an operational setting: (1) Aurora project 2001–2004, Temperature Sensor Accuracy, and (2) an Aurora-initiated project that has now developed into National Cooperative Highway Research Program Project 6-15, Testing and Calibration Methods for RWIS Sensors. Although both of these projects represent a move in the right direction, pavement sensing is a continuously developing field, and a structured test and evaluation procedure needs to be developed and administered on an ongoing basis.

In addition to the need for improved testing of existing environmental sensors, improved sensor technology specific to road weather applications must be developed. In particular, environmental sensors must be robust, reliable, and accurate under a wide variety of weather conditions. Research is needed to develop improved techniques for measuring quantities of special relevance to road weather applications, such as the amount of ice, precipitation, and chemical treatment on the road surface.

Of particular concern is that current pavement sensors are limited in their spatial representativeness, in their ability to function in all moisture conditions, and in the assumptions made in signal processing. Presently there are several different pavement sensor technologies in use or being developed. Some of these include fiber optics, active sensors, resistance, spectral cameras, and even analysis of photos from still and video cameras. While each of these methods has merit, they all have weaknesses that must be more thoroughly understood in a field environment; for example, sensors measuring chemical concentration on pavement function with a reasonable degree of accuracy only when there is plenty of moisture on the road surface. Under dry conditions passive sensors cannot distinguish different types of chemicals or surface contaminants, and active sensors change the environment surrounding them, sometimes enough to affect these measurements. To be truly effective in making equipment control decisions during frost or other low moisture conditions, pavement sensors need the ability to measure when little or no moisture is present.

Besides changes in sensor technology, research is needed to improve information processing capabilities in environmental sensors. In most cases raw information collected from a sensor undergoes some processing in the field before it is reported, however the processor technology does not seem to have advanced far enough to fully meet user needs at this time. Pavement sensors hold great promise for significantly expanding the spatial coverage of observations and thereby act as an important constraint in models of road conditions; however, these instruments may require substantial research and development to attain required levels of accuracy and reliability.

Recommendation: Improve road weather instrumentation.

The road weather research program should support efforts to establish science-based standards for road weather sensors and to implement procedures to test, compare, and certify that sensors meet these standards. The comparison and testing procedures need to ensure that the results are applicable to a field setting. The committee recommends support for the development of improved sensor technology specific to road weather applications, such as for the amount of ice, all-season precipitation, present weather, and chemical composition of the slurry on the road surface. Research is needed on design and deployment of sensor systems to provide data more generally representative of the traveled road surface and that are reliable under a wide variety of weather and traffic conditions.

Data Quality Control, Dissemination, and Archiving

Systematic quality assurance, dissemination, and archiving of road weather data are fundamental steps to making progress on all the recommendations made here. Some types of data must be collected at the time of road construction (e.g., soil type and road construction materials) and archived as a semipermanent database, and updated during renovations. Other forms of data (e.g., change of land use along the roadway) need periodic updating within the database. On the other hand, collection of real-time data leads to a very dynamic, ever changing database. The utility of road weather data would be enhanced if the data were easily accessible in common data formats and geolocated to the extent possible (Reichardt, 2003). Many state departments of transportation are already taking advantage of GIS in storing road composition data, but this is not yet standard practice in the meteorological community (Wilhelmi and Brunskill, 2003). An encouraging development is the NWS designing the National Digital Forecast Database to be GIS compatible. This will greatly increase the value of these forecasts for both direct use and private sector partners.

Quality assurance is essential for data reliability and accuracy. Much of quality assurance can be handled automatically, with checks performed continuously on the data stream. Periodic automatic scans of datasets using statistical tools can identify longer-term trends indicative of instrument drift; however, careful visual scrutiny of data displayed either as time-series graphs or as two-dimensional data fields by an experienced analyst is still an essential part of quality assurance. Quality control supported by field maintenance teams and calibration laboratories is a necessary extension of the quality assurance program.

A dissemination program ensures that data necessary for research or operations are provided when and where needed. Often research programs can be satisfied by providing researchers direct access to the database (a "pull" situation), whereas operations often require that specified data be sent to them (a "push" operation). Data portals, providing easy access to data stored in a variety of formats, often at different locations, are a growing trend but require significant investment in professional time and technology to establish. A particularly demanding criterion in providing real-time data is latency, or the time lapse between taking a measurement and providing the resulting value to the user. Some applications (e.g., warning of severe weather events) have latency demands of a few seconds.

There are many state, regional, and national programs undertaking the challenge of merging available observations and disseminating the information in a timely, meaningful, and easily interpretable format. One such program is the Meteorological Assimilation Data Ingest System operated by the National Oceanic and Atmospheric Association's (NOAA) Forecast Systems Laboratory. This system collects surface, profiler, upper air, and aviation data (Figure 4-1) from meteorological networks, both public and private, for North and Central America (Miller and Barth, 2003). Data are processed continuously at five-minute intervals and can be acquired by interested user groups over the Web (http://jailbird.fsl. noaa.gov/MADIS/). A new proposal, the Sensor Web (Reichardt, 2003), provides an opportunity to collect and make available multisource, formatted sensor data to scientists, traffic managers, disaster managers, and others. This appears to be a viable opportunity to collect the high volume and varied types of data that are or will be available.

Archiving of data requires not only storage in readily accessible form of the data itself, but also of the accompanying metadata. Metadata provide the history of data and are a dynamic quantity in their own right. Records can contain both descriptions of the data formats and a wide range of information, such as sensor calibration history, exposure imagery, and records on how frequently the data are used and by whom. Management of a data archive is a technical challenge as the data must be secured against physical hazards and unauthorized changes but still be readily accessible. An additional archival challenge is forward migration of large databases as the storage medium is upgraded.

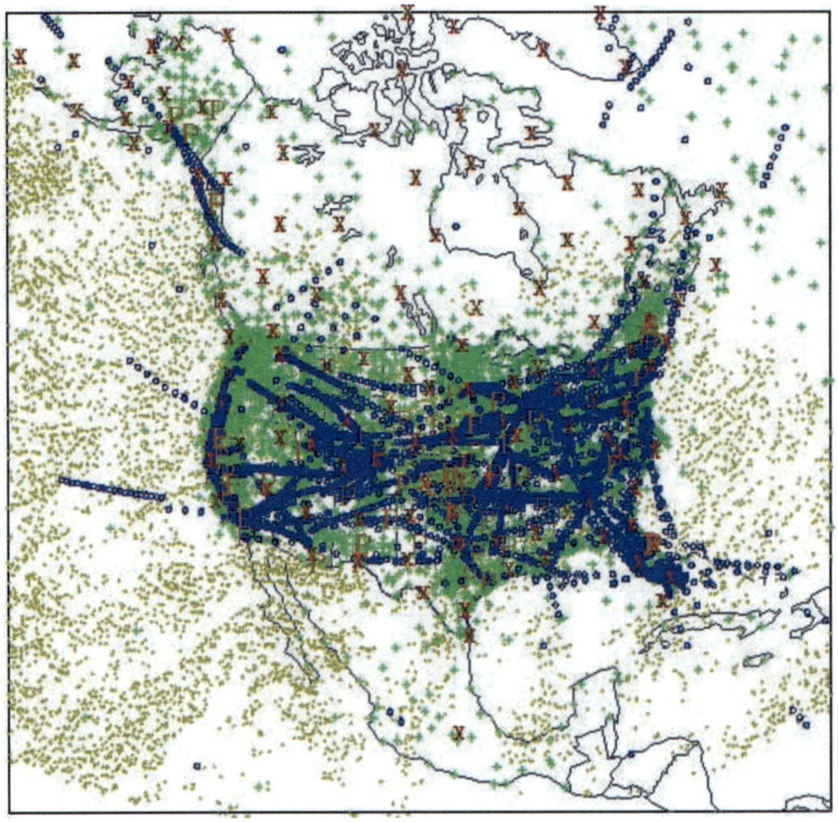

FIGURE 4-1 Meteorological Assimilation Data Ingest System observations available in North America. Green represents surface data, blue represents aircraft data, red represents radiosonde data, and the yellow "P" represents profiler data. SOURCE: NOAA's Forecast Systems Laboratory.

Recommendation: Maximize utility and quality of road weather information.

The committee recommends that the road weather research program take steps to maximize the utility and quality of road weather information. First, to facilitate sharing among multiple entities, the road weather community should help standardize formats of geolocated data and associated metadata (e.g., the National Transportation Communications for Intelligent Transportation System Protocol for Environmental Sensor

Stations). Second, the community should standardize the presentation of road weather information (e.g., in terms of units, colors, or symbols) to user groups across North America. Third, to facilitate integration of information from different research communities, the community should use geographic information systems to analyze, manipulate, and display road weather observations, model output, and forecasts. Fourth, the community should establish standards and implement data quality assurance and control procedures. It is critical to ensure that all road weather observations are of sufficient quality to be used in the numerous desired data applications.

Observations and Infrastructure for Validating Models

Specific observational data also are needed to validate and improve both road and atmospheric boundary layer models. In addition to exploiting untapped data sources and expanding current observational capabilities, separate sustained support is needed for establishing an infrastructure to obtain observations specifically for both model and road weather forecast validation. Ideally as many validation systems as possible should be deployed, but at a minimum they should be located along the national demonstration corridors (see Chapter 5, "Establish National Demonstration Corridors") and possibly in conjunction with regional research centers (see Chapter 5, "Establish Regional Research Centers").

To capture the diverse roadway environment that exists across the nation, the observational facilities should be located along roads with various types of land cover, terrain, and uses. In some cases these facilities should consist of permanently instrumented arrays in key areas that should be routinely maintained and operated. These core networks should provide the long-term data records necessary to investigate a number of road weather challenges. In addition to the permanent arrays, instrumentation is needed for rapid deployment for occasional intensive observation periods. To be cost-effective the deployments should likely occur in and be complementary to the permanent arrays. The platforms should cover a sufficient area and depth to be representative of both the roadway and the environment on either side. Where possible the facilities also should be augmented by existing mesonet facilities to obtain even more data and further increase cost-effectiveness.

The most important parameters that should be recorded at all observational facilities are temperature, precipitation amount and type, wind speed and direction, relative humidity, insolation, pavement temperature, road surface cover, and surrounding surface cover. The data from these observational facilities should be of sufficient temporal resolution (e.g., data collection as frequently as every minute) to provide a sufficient amount of data and to capture rapid environmental changes. These specialized arrays will facilitate the evaluation and improvement of road weather models and forecasts.

Recommendation: Develop observing capabilities to measure the performance of road weather forecasts.

The committee recommends that the road weather research program develop special observing facilities to validate forecast tools and verify road weather forecasts. Ideally, these observing facilities would be located along the national corridors and at the regional centers identified previously. In some cases these facilities could consist of permanently instrumented arrays, but to be cost-effective most should consist of supporting infrastructure that facilitates the rapid deployment of instrumentation for occasional intensive observation periods. An important aspect of the recommended work would be the collection of micrometeorological measurements necessary to develop, improve, and validate the models.

ADVANCING THE SCIENCE AND DEVELOPING THE TECHNOLOGY

The national road weather research program will need to coordinate and support efforts to increase understanding of road weather phenomena and options for increasing safety, mobility, and efficiency of the nation's roadways during inclement and other high-impact weather. Because little research has been conducted specifically on these issues to date, targeted support of research on the road weather environment could potentially yield substantial returns in understanding and consequently managing the roadway system. Areas of meteorology where increased attention to the roadway environment is needed include boundary layer meteorology; weather-influenced phenomena that impact roadway transportation, such as flash floods and fires; and the development of a reliable winter severity index. Several aspects of transportation research would benefit from increased attention to weather, such as quantifying how weather affects traffic, human factors associated with driver behavior during inclement weather, and improvements in vehicle technology to better equip drivers to respond to weather conditions. These topics, discussed in more detail in the following sections, are those that the committee believes require increased attention initially; they do not, however, represent an exhaustive list of basic research needs. Indeed, as the research program is established and grows, the research must evolve to meet changing needs.

Meteorological Phenomena That Affect the Roadway

Micrometeorological processes that occur near the bottom of the atmospheric boundary layer, which is the layer of air that directly feels the effects of the ground surface, can be quite complicated and make it difficult to analyze weather along roadways. On sunny days the boundary layer is roughly 1 to 2 km deep and is formed by turbulent mixing of air currents rising from the warm surface and compensating downdrafts. At night the boundary layer is much shallower. During both day and night, vertical gradients in wind, temperature, and humidity are concentrated near the surface, and observations of atmospheric conditions even 2 to 5 m above the surface may not capture the conditions that affect driving and the road surface. To understand the effects of the surface on the atmosphere, particularly at scales relevant to roadways, micrometeorologists often must account for the effects of canopy, terrain, natural and built structures, surface vegetation, and soil characteristics (see Figure 4-2); for example, the wetness of the ground affects the partition of solar energy into evaporation and heating, whereas natural and built structures can alter heating and cooling through radiative processes and can deflect or channel the wind.

Significant research challenges remain in understanding and predicting micrometeorological influences on temperature in the roadway environment. Nocturnal cooling, for example, can strongly affect the roadway environment. On nights with light wind and clear skies, the ground cools rapidly by radiating heat to space and then cools the shallow layer of air immediately above by contact. On a slope, this cold air nearest the ground slides downhill because it is heavier than the air at the same height further out into the valley. These down-slope winds are reinforced by flow from the valley sides and continue down the valley, eventually forming pools of cold air in the valley floor (Figure 4-3). Even in gentle valleys, air can be as much as 10°C cooler than the air on the ridges. Cooling in depressions protected from the wind and wind-induced mixing ("frost hollows") can be even more significant. In both cases the low-lying areas will be most susceptible to fog or frost.

Nearby obstacles and terrain further increase the complexity of the road weather environment. Trees, bluffs, bridges, and buildings can deflect and

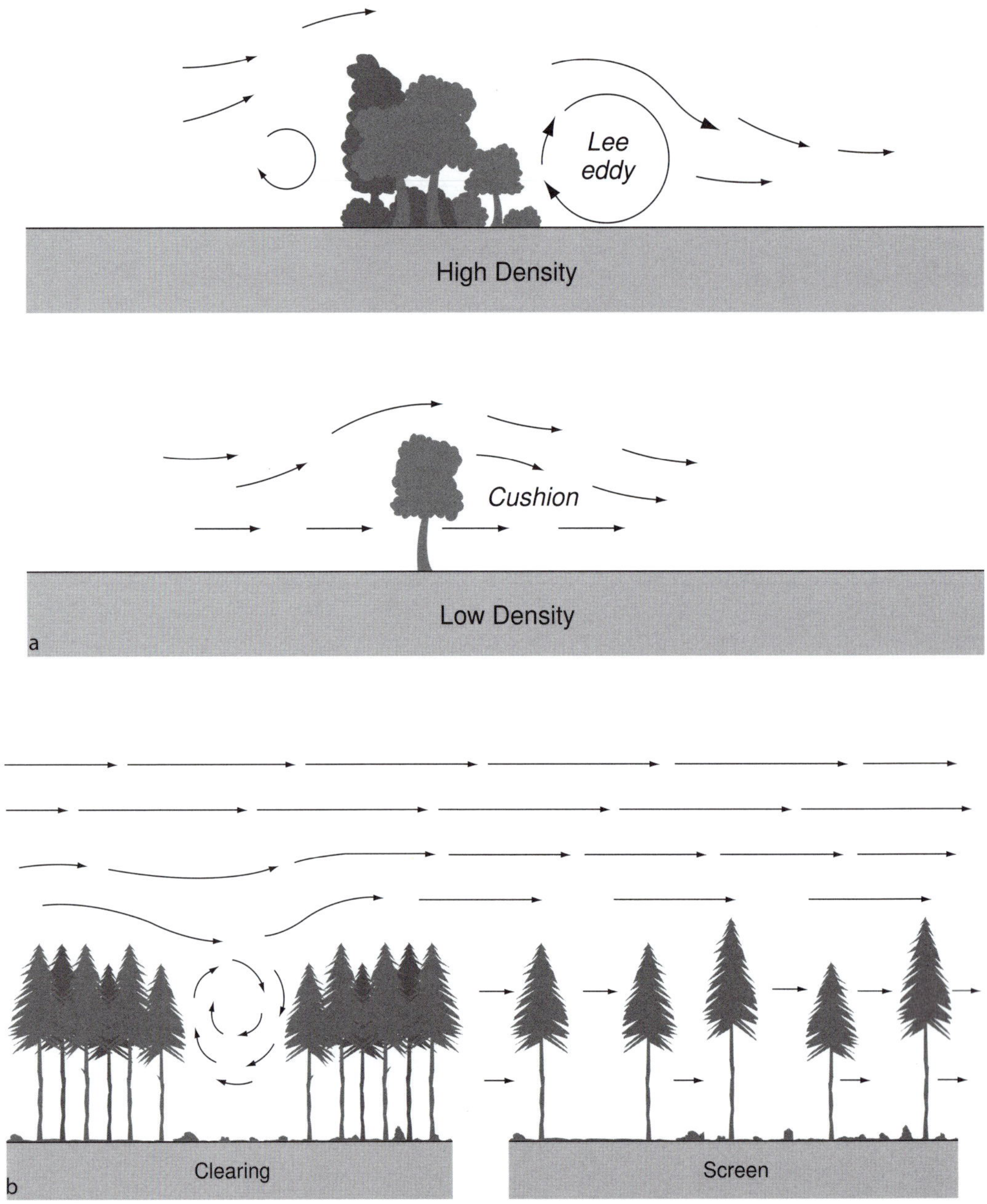

FIGURE 4-2 (a) Wind speed reduction in the vicinity of shelter belts with different densities. (b) Airflow in a regeneration area and under a screen of old trees. SOURCE: Oke (1987).

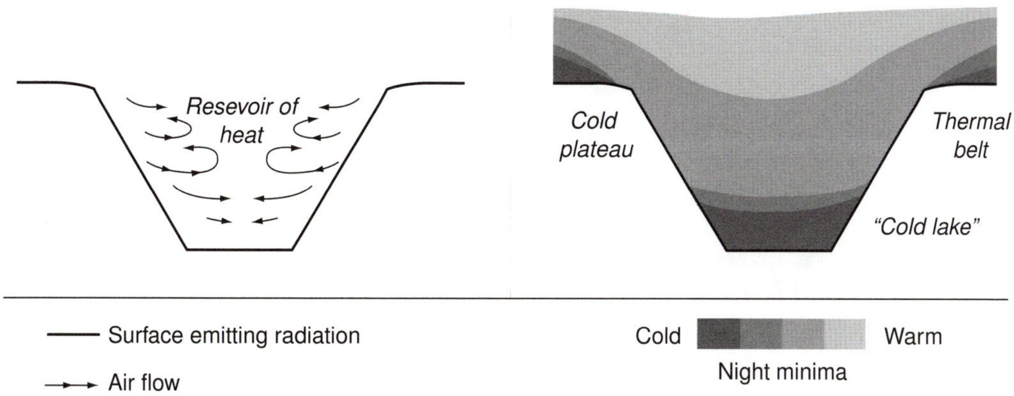

FIGURE 4-3 Development of a frost hollow in a depression. SOURCE: Geiger et al. (1995).

funnel wind, potentially causing drifting snow and unpredictable, hazardous gusts. Heat exchange between the road surface and the atmosphere is complicated by the heterogeneity of the road surface and its nearby environment. This environment is complicated further by the traffic itself, which introduces heat, mixes air, splashes fluid on the road, and reduces the radiative warming and cooling of the road surface.

The importance of local effects to the roadway environment suggests a two-step approach to simulating road weather conditions: (1) running atmospheric models with land surface data (except for ESS data from stations sited to document nonrepresentative effects) to solve the future state of the atmosphere at scales down to 5 to 10 km; and (2) using that future state of the atmosphere along with the road information (pavement and near-road environmental sensors) to produce road surface temperature and road condition forecasts. For roads in flat terrain and fairly uniform environments, idealized models that characterize the exchanges of heat, moisture, or momentum between the air and the surface can be useful. For roads in complex terrain one must resort to higher-resolution atmospheric models with detailed surface characteristics or to empirical techniques to successfully capture the fine-scale variability.

Another micrometeorological research challenge is the observation and forecast of conditions that reduce visibility, such as fog or blowing snow. Although many fog-prone stretches of road are known locally, there has been relatively little research on the fog problem specific to road transportation. No one has succeeded in the operational modeling of the development of radiation fog. Similarly, enough is known about how snow accumulates in certain areas to install snow fences to mitigate snow accumulation on the road, but less attention has been paid to loss of visibility from falling or blowing snow, or slush splashed on windshields by passing traffic.

Research is needed to improve visibility detectors. Currently available visibility sensor equipment takes samples over a very short distance or baseline (1 to 3 feet) and assumes this to be representative of the entire area. Clearly this assumption is not always valid because visibility in fog, snow, blowing dust, and other inclement weather conditions, tends to vary greatly over short distances. One possible means of achieving a more representative measurement is through analysis of video frames or still images. While there has been some research done in this area, the technology has not been commercialized and is not available for use at this time. Care must be taken to seek instruments that measure low visibility specifically under all lighting conditions.

Weather-Influenced Phenomena

Many weather-induced hazards, such as flash floods, fires, avalanches, snowdrifts, rock falls, mudslides, and other debris flows, rapidly impede roadway use. Many of these hazards are in regions with sharp regional and local gradients in weather and climate behavior; for example, an avalanche may flow over several thousand feet of vertical distance before reaching a roadway and then affect only a small length of the road where it crosses. The region in which the avalanche began may have completely different weather conditions than where the avalanche covers the roadway (Figure 4-4). Thus, not just the roadway conditions require monitoring and forecasting but all of the regions that either directly or indirectly affect the roadway. Research is needed to improve both observing and forecasting these hazards and how best to respond to them.

Observational, forecasting, and emergency response capabilities vary from region to region. Most NWS offices have service hydrologists on staff; many counties, towns, and cities have active search and rescue teams; some regions of the country have avalanche prediction, research, and warning centers. Many universities, private companies, and government agencies are investigating weather-influenced hazards with new and improved simulation and modeling research, improved sensing and measurement technologies, and merging of hydrometeorological data. To effectively implement this research for use in the roadway environment requires (1) an

FIGURE 4-4 Removal of avalanche debris along US-2 in Washington. SOURCE: Northwest Weather and Avalanche Center, Seattle, Washington.

operational focus; (2) information that is easy to implement and use; and (3) weather liaisons between the research and development entities and the practical application by users.

A number of significant gaps should be addressed in order to improve the forecasting and warning of these weather-related roadway hazards. First, research is needed to improve the understanding of the causes of these hazards. Second, efforts are needed to (1) improve sensing and measurement technologies (e.g., snow depth, precipitation, mountaintop winds); (2) develop the ability to combine various types of hydrometeorological data; and (3) facilitate sharing of real-time data and products among entities. Using a GIS-based system, "smart maps" should be created for hazard regions across the country that contain information such as roadways, potential weather-related risks, and weather information. Eventually, improved simulation and modeling of these weather-influenced incidents can be used to develop improved site-specific forecasts for weather-induced hazards that could be incorporated into these smart maps. Finally, emergency responses to these events require proper coordination, as many agencies may be involved.

Recommendation: Improve prediction and warning of weather-influenced hazards that rapidly impede roadway use.

The road weather research program should support research to improve predictions and warnings of weather-induced hazards that rapidly impede roadway use, such as flash floods, avalanches, mudslides, and other debris flows. Better understanding of the causes of these hazards and how best to respond to them could help avoid loss of life and property. Existing observational and forecasting capabilities need to be assessed and significant gaps should be addressed.

Winter Severity Index

Winter weather indexes have been developed in several states as a means to normalize the classification of weather events and the environment in which they occur. Similar indexes have been developed for the accumulated effect of storms through an extended period of time or season. One such winter weather index is a multipurpose tool that determines the relative severity of winters and enables practitioners to determine the efficacy of their responsive operations. It can be used to (1) estimate winter maintenance costs for improved budgeting and resource planning; (2) measure outcome in terms of regain time (i.e., the rapidity of the clearing of roadways); (3) compare different types of operations to determine which practices are best; (4) compare new equipment, chemicals, and technologies for improving road-clearing efficiencies; and (5) develop modeling and sophisticated scenarios to optimize strategies. These indexes can be developed locally, for instance, to develop best and worst case budgeting scenarios; or they can be developed for interregional use to compare severity of winter weather conditions.

Depending on its purpose, the components of a winter severity index can vary greatly, but the types of variables used fall into the three main categories of geography, solar input, and weather conditions. Geography accounts for elevation, latitude, longitude, topography, and road orientation. Solar input is based on cloud cover, the surrounding environment (e.g., trees, rocks), and the time of year, which influences both temperature and ground cover. The wind speed and direction, temperature, and precipitation type, quantity, and duration are among the important weather conditions.

An example of a winter severity index—the Hulme index—is given in the following equation. This index has been used to evaluate interannual variability of salt use and other winter road maintenance costs, and to assess the cost-effectiveness of environmental sensors.

$$WI = 10T - F - (18.5S)^{\frac{1}{3}} \pm C$$

In this equation WI is the winter index value, T is the mean maximum air temperature in degrees Celsius, F is the total number of nights with frost on the ground, S is the total number of days with snow cover at 9:00 a.m. local time, and C is a constant such that the climate averaged WI equals zero. Negative values of this index imply more severe winter conditions than average, and consequently are associated with greater road maintenance costs (Andrey et al., 2003a).

Relatively little research has been devoted to devising optimum winter weather indexes, even though many road weather maintenance operations depend on such indexes to evaluate their operations and to make the case to state and local governments for additional funding. Research is needed to improve the formulation and standardization of winter weather indexes. ESS data and other localized in situ weather measurements offer an opportunity to automate a winter weather index with objective, fine-scale measurements.

Weather Impacts on Traffic

Despite a general appreciation among transportation practitioners that weather has significant effects on traffic, little observational data are available to quantify these effects. Data routinely collected from vehicle detectors and video cameras could be used by researchers to better understand the relationships between weather and traffic. For the most part these data, which are collected by traffic operations centers in some states and cities, are used for real-time management of the transportation system but are not archived. There is an opportunity to learn from these data about impacts of weather on traffic flow and to refine estimates of traffic simulation parameters; for example, comparing traffic conditions on days when inclement weather took place with those on fair weather days could help constrain parameters used in traffic simulation models. Taking better advantage of these traffic data for research purposes will require finding ways to archive research-quality data cost-effectively.

Recommendation: Archive and mine operational traffic observations to assess weather impacts.

Some states and cities monitor traffic throughout their area and routinely collect data from traffic detectors, video cameras, and other sensors. Often these data are used for real-time management of the transportation system but are not quality assured or archived in formats or in locations readily accessible to the research community. The committee recommends that the road weather research program take better advantage of these data, as they provide many opportunities to learn about impacts of weather on traffic flow and refine estimates of traffic simulation parameters. This will require finding cost-effective ways to quality assure these data, archive them in easy-to-use formats, and make them accessible to researchers.

Human Factors

The discipline of transportation human factors addresses a wide range of human responses to driving conditions, vehicle behavior, traffic, and information intended to assist the driver. A number of factors related to driver behavior affect decision making, particularly during periods of stress. Some of these factors are (Badger, 1996)

- *Experience:* Time and experience behind the wheel in all types of weather conditions
- *Familiarity:* The extent to which the driver is familiar with the vehicle and roadway characteristics
- *Memory:* The ability to recall previous warnings and appropriate action
- *Vision:* The ability to discriminate highway features, traffic, and weather and roadway conditions
- *Perception:* The way people view the weather conditions, their severity, and impact
- *Information processing:* The willingness to access available information about driving conditions and the ability to comprehend and respond to information appropriately
- *Driver response time:* The ability to react to changing conditions, warning signs, and weather information
- *Fatigue:* Fatigue adds to an already complex set of circumstances when adverse weather enters the picture

- *Alcohol, drugs, medications:* Similar problem to fatigue
- *Age:* Older drivers have slower reactions but younger drivers are more impulsive drivers
- *Neuropsychological, medical, psychiatric disorders:* These problems compound response time and comprehension of weather information

Given the wide variety of factors affecting driver decision making, understanding the variability in driver behavior in adverse weather conditions is particularly challenging. Additional complexity is introduced by the way weather information is transmitted (e.g., radio, dynamic message signs, in-car display), received, and processed.

Research on human factors considers the type of information that should be provided to drivers to elicit a desired response and how drivers will use and respond to new technologies and information. Current research is beginning to consider the relative value of in-vehicle and roadside dissemination of weather information, the type of information that should be provided to drivers, and how the urgency of the information should affect presentation (Collins et al., 1999). For example, researchers are experimenting with visual, auditory, and haptic methods (e.g., a pedal vibration when antilock brakes engage) for communicating to drivers. Another research emphasis is on how to increase visibility during inclement weather through innovative visual enhancement systems or pavement markings. Researchers at Virginia Polytechnical Institute and State University have investigated the ability of high-output halogen, bending light, motorway light, ultraviolet A, and infrared radiation in enhancing the ability of motorists to visualize objects through rain, fog, and other conditions that degrade visibility (Chrysler et al., 1997; Gish et al., 1999; Nilson and Alm, 1996; Olson et al., 1990).

Despite these initial efforts little research exists that relates human factors and weather situations. Not well understood are such simple things as reaction to weather information and highway condition messages on highway dynamic message signs and severe weather warnings on the car radio. Additional research is needed to understand how driver behavior changes when new tools or capabilities are made available to them; for instance, do they become overconfident in their ability to drive in inclement weather or do in-vehicle display or communication systems distract the driver? The same applies to human factors in highway maintenance and engineering.

Weather and roadway research must consider human factors in effective implementation of results. The Federal Highway Administration (FHWA), through its Advanced Traveler Information System (ATIS) program, has acknowledged that while technology exists to display information to the driving public, there is little knowledge to guide the proper selection of methods and displays. The agency knows that key factors in such selection include "accessibility, legibility, and understandability of ATIS information, the potential for ATIS information to facilitate driver decision-making, and the potential for ATIS information to distract the driver from the primary task of controlling the vehicle" (http://www.fhwa.dot.gov/tfhrc/safety/pubs/96147/sec4/body_sec4_04. html). Figure 4-5 captures this concern. Road weather research that considers the human factors in driver decision making will help frame the best methods and formats by which information is transmitted to the driving public.

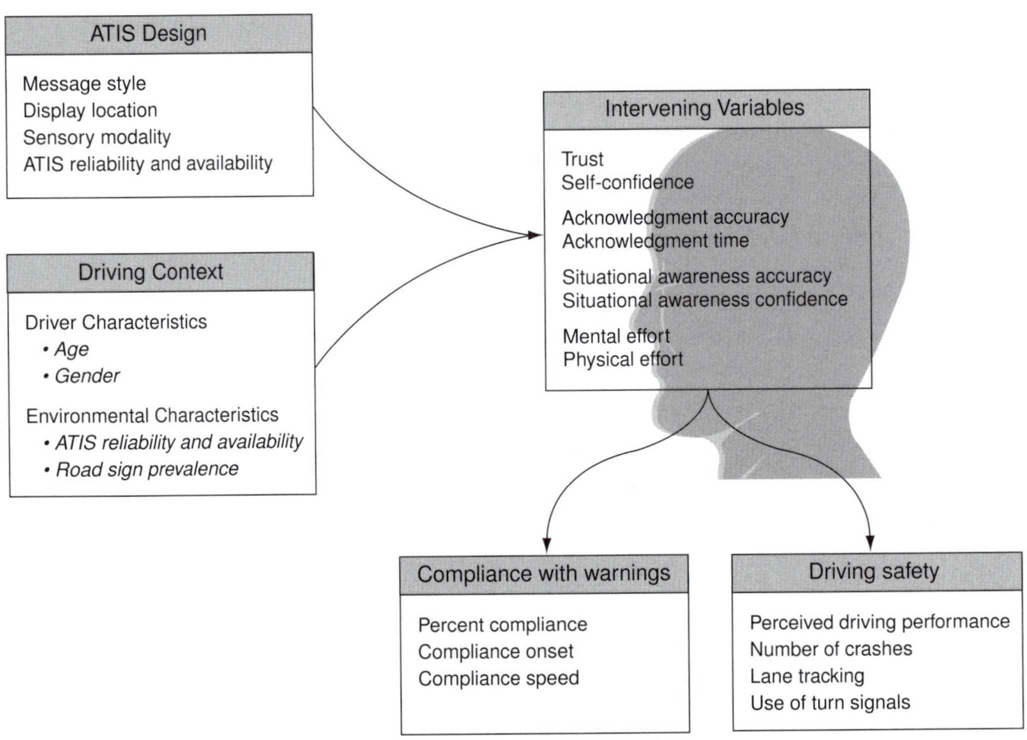

FIGURE 4-5 Factors moderating the effect of ATIS design characteristics on driving safety and warning compliance. SOURCE: Federal Highway Administration.

Recommendation: Support human factors research to obtain desired responses to road weather information.

The road weather research program should support research into the human and cultural factors involved in driver decision making under a variety of weather conditions and in driver response to new technologies for the delivery of road weather information. Important aspects of this research include the information, content, and format for optimum response, and safe methods for communicating with vehicle operators, especially during hazardous weather conditions. The FHWA's program for an Advanced Traveler Information System has acknowledged that technology exists to display information to the driving public, but there is little knowledge to guide the proper selection of methods and displays. This research should build on the work done by the Advanced Traveler Information System, Commercial Vehicle Operations, and Advanced Traffic Management Systems and be compliant with the National ITS Architecture Maintenance and Construction Operations user service bundle. Additional research is needed on effective communication of road weather information to traffic managers, maintenance personnel, and emergency managers, as they have specialized needs that are significantly different from those of the general public.

Vehicle Technology

Major advances in vehicle technologies for providing drivers with improved weather information have come about through research and development of "telematics," or in-vehicle information and communication technologies and services. The private sector has led research and development efforts in the field of telematics to date. A variety of telematics features are available or under development, including

- safety, security, and location services;
- remote vehicle functions (e.g., door unlock, sound horn);
- vehicle maintenance diagnostics and prognostics;
- personal information services (e.g., stocks, news, weather, traffic, text-to-speech e-mail); and
- hands-free phone with voice-activated calling.

Several of these technologies require a capability for a vehicle to communicate directly to a central server where data processing takes place and appropriate responses are set into motion. A centralized repository and server is also needed for providing information to vehicles and drivers. In some cases this central server may be operated by the automobile manufacturers, while in others a third-party vendor may be involved in retrieving and processing data from automobiles and responding to driver requests for information.

Telematics systems require technology in the vehicle, a wireless carrier of voice and data (e.g., cellular), and a service delivery network. Typical technologies used on the vehicle include the global positioning system, communication capabilities (i.e., audio interface, microphone, antenna), and hardware and software to implement instructions to the vehicle. Either embedded telecommunication units or portable telephones, or both, are used to communicate between the vehicle and the service delivery network. The embedded units have the advantage of always being "on" but are less portable than portable phones. Using both an embedded unit and a portable phone provides the highest level of functionality but is more expensive. The service delivery network includes databases that store information of potential interest to motorists and those that store information about performance and possible maintenance needs downloaded from the vehicle. The service delivery network may also include a workforce to respond individually to driver requests.

Ford Motor Company estimates that there will be 12 million telematics service users by 2008 (Miller, 2003). Even so, many original equipment manufacturers are stepping back from major telematics commitments as they reevaluate the business model to determine which services customers want and how much they will pay for them. In the United States, emergency roadside assistance has been the most widely adopted. The OnStar system is the most extensively used telematic service in the United States, with responses monthly to 500 air-bag deployments, 15,000 remote-door unlocks, 375 stolen-vehicle tracking calls, 200,000 route-support calls, and 14,250 road assistance calls (http://media.gm.com/division/2003_prodinfo/03_corporate/telematics.html). European telematic systems that include navigational services in addition to roadside assistance have gained popularity, while the Japanese have more quickly adopted both services and Internet-based infotainment.

The development of "smart vehicles" is a critical component of intelligent transportation systems. Intelligent vehicle systems can complement the information gathered by highly instrumented roadways. Some examples of technologies under development include precrash sensing using radar and optical sensors along with wireless communication between vehicles; sensing of vehicle occupants so as to optimize vehicle performance; instruments to assess the surrounding environment, including surface friction, temperature, and precipitation; reconfigurable and "head-up" displays; and automatically adjusting such features as windshield wiper function and the temperature of the car interior. Of particular relevance to weather is smart navigation, in which sensors on the vehicle report road conditions (e.g., iciness, traffic, incidents, construction) back to a traffic operations center, and the server sends to all drivers on that road alternate routes as necessary. Further research is needed for the roadway community to take full advantage of these new vehicle technologies.

Part of the vision for the new generation of vehicles is that they should communicate with each other about their speed, location, and heading information. This capability, combined with omnidirectional sensors that would alert the driver to potential collisions, could dramatically decrease the likelihood of accidents. Sensors in the vehicle that identify poor road conditions (e.g., iciness) could communicate the location and problem to vehicles traveling behind it. This would allow the driver or the vehicle to adjust to the conditions.

Most original equipment manufacturers are looking to provide new telematics technologies with an open architecture. They are not planning to be service providers and will not develop content or the service delivery network. Rather, the open architecture will allow other businesses to enter that market. Several manufacturers have agreed to standardize the interfaces between vehicles and communication networks. They have founded the Automotive Multimedia Interface Collaboration to facilitate the development and standardization of multimedia interfaces to vehicle communication networks. For these standards to be of optimum utility they should have an open architecture and be integrated into ongoing intelligent transportation systems (ITS) development.

MODELING CAPABILITIES AND FORECAST TOOLS

Improving modeling capabilities and forecasting tools should be a key objective of a new road weather research program. These efforts should build on existing capabilities in both the meteorology and transportation communities and exploit potential new capabilities (Table 4-2). As discussed in more detail in Chapter 3, numerical weather prediction has increased substantially in sophistication and performance since its inception in the 1950s. However, these efforts have largely centered on forecasting synoptic and mesoscale weather conditions on scales larger than the spatial

TABLE 4-2 Existing and potential models that should be applied to road weather research and operations

Model Function	Model	Description
Numerical weather prediction	Eta	Highest-resolution National Centers for Environmental Prediction model with a 12-km grid spacing. All of North America is included in domain with forecasts to 84 hours. Model output includes conventional fields such as pressure, precipitation, dewpoint temperature, temperature, heights, and winds, plus many other derived quantities.
	Regional mesoscale models, including, for example, Regional Atmospheric Modeling System (RAMS) and Fifth-Generation National Center for Atmospheric Research (NCAR)/Penn State Mesoscale Model (MM5).	Fine horizontal scale (grid spacing of a few km) with grids usually multiple nested. Lateral boundary conditions from larger-scale operational models.
	Weather Research and Forecasting (WRF)	High-resolution research and forecast National Weather Service model with horizontal grid of 1 to 10 km. A new community model being jointly developed. Plans include assimilation of very-high-resolution data such as radar.
Modeling the roadway environment	One-dimensional energy-balance models	Models vertical exchanges of heat and moisture within a vertical column by including weather information and road surface properties
	Multi-dimensional energy balance models	Models full three-dimensional energy balance by accounting for the complex terrain of an environment (e.g., vegetation, topography)
Traffic simulation and management	Dynamic Network Assignment-Simulation Model for Advanced Road Telematics (DYNASMART-X)	Mesoscale (areawide) planning model that provides estimates of traffic conditions, flow patterns, and rerouting based on simulated driver choices

Operated by	Road Weather Application	Limitations
• National Weather Service • Workstation version available to private sector and academia	• General weather outlook through day three • Subsynoptic-scale weather prediction • Highway maintenance decision making • Can provide initial conditions and boundary conditions for road-surface models	• Insufficient resolution to capture finer mesoscale phenomena (e.g., thunderstorms, small terrain-influenced circulations) • Much too coarse to provide direct forecast for roadway environment • Inadequacy of data to define initial atmospheric conditions and the surface boundary conditions (e.g., landscape, soil, and vegetative conditions) • Inherent limits of predictability on the desired time and space scales
• Academia • National Weather Service local partners • Private sector	• Mesoscale weather predictions specific to a certain region • Highway maintenance decision making • Can provide initial conditions and boundary conditions for road-surface models	• Incomplete or lack of coverage for parts of United States • Intermittent operations • Vulnerable to outages • Unsteady funding
• Government labs • Academia • U.S. Air Force	Mesoscale weather prediction	• Relatively new model still in development • Not currently available for routine operations • Development research is ongoing
• Academia • Private sector • Government labs	Predicts the temperature and wetness of a roadway surface for tactical (< 6 hours) and strategic (> 6 hours) purposes	• Only completely accurate when atmosphere and road surface conditions are horizontally homogenous
• Academia • Private sector • Government labs	Predicts the temperature and wetness of road surfaces and their surrounding environment	• Computationally intensive • Expensive • Requires substantial input data
• University of Maryland • University of Texas	Traffic management in adverse weather	• Difficult to model human behavior • Relies on good time-varying origin-destination data

continued

TABLE 4-2 Continued

Model Function	Model	Description
Traffic simulation and management (continued)	Dynamic Network Assignment for the Management of Information to Travelers (DynaMIT)	Mesoscale (areawide) planning model that integrates historical databases with real-time data to estimate traffic conditions, predict future conditions, and provide information to drivers
	Hurricane evacuation models	Models that simulate roadway operations (e.g., congestion, clearance times) using weather, flood, traffic, and geographic data to determine best routing and re-routing decisions
Decision Support Systems	Maintenance Decision Support System	Assembles weather conditions of an area, predicts the potential for deteriorating conditions, plans treatment scenarios, and provides decision makers with treatment recommendations
	High Performance Paving (HIPERPAV)	Prediction tool that uses temperature, wind, humidity, and cloud cover information to predict optimum paving windows for roadway construction
Coupled atmospheric-traffic simulation models		End-to-end models that integrate weather and traffic data to guide decision support

scale of a roadway. Models of traffic and road surface conditions are available, though they are not used as routinely as are weather models.

The road weather system of the future will depend on greatly improved forecast capabilities, but attaining the desired skill in forecasting weather and road conditions on the scale of the roadway will require significant advances in modeling and empirical techniques. For example, current numerical weather prediction capabilities, described in more detail in Chapter 3, are limited in several ways, including

- inadequacy of data to define initial atmospheric and surface boundary conditions (e.g., landscape, soil, and vegetative conditions);

Operated by	Road Weather Application	Limitations
Massachusetts Institute of Technology	• Traffic management in adverse weather	• Difficult to model human behavior Relies on good time varying origin-destination data
• Academia • Private sector • Government labs	• Traffic management during evacuations • Can be used in real-time by emergency management officials • Locate emergency shelters for drivers	Real-time models require substantial data input
Federal Highway Administration	Provides maps of road weather conditions along road corridors for 1 to 48 hours for maintenance crews	Relies on the accuracy of weather forecasts
The Transtec Group	Optimal roadway construction based on weather conditions	Relies on accuracy of weather input data
Not yet developed	• Improved traffic management in adverse weather • Simple, more uniform information for users and maintainers	• Computationally intensive • Requires substantial input data

- suboptimal resolution and accuracy of numerical weather prediction or empirical models; and
- inherent limits of predictability on the desired time and space scales.

Advances in future roadway weather forecasts will depend on how well the high-resolution land surface properties (e.g., soil moisture, topography, vegetation type, soil type), hydrological aspects (e.g., soil infiltration, run-off), cloud microphysics, and boundary layer physics are parameterized and incorporated into numerical weather prediction models. As shown in Table 4-2, other modeling approaches applied to the road weather problem are similarly limited, for example, by computational capabilities or difficulty in representing human choices or for institutional reasons. How to support the development of these models and other forecasting tools strategically and cost-effectively will be a challenge facing the proposed road weather research program.

Deficiencies in the parameterizations within the numerical weather prediction model are frequently responsible for significant biases in predicted parameters especially within the boundary layer, which is the most critical to road weather forecasts. As such, post-processing of model output is often necessary. Historically, this has been done subjectively by the human forecaster with the aid of statistical tools and experience. In addition, the forecaster has routinely added value by "smoothing" run-to-run inconsistencies and offering expert interpretation of the objective predictions.

The NWS has just started to present its forecasts digitally through the Interactive Forecast Preparation System and the National Digital Forecast Database. This system is designed to take advantage of the resolution of the prediction models, yet allow the forecasters to continue to play a significant role in the process. The forecasts are delivered on a 5-km grid through seven days and deliver many forecast parameters. The National Digital Forecast Database system should provide basic high-resolution forecasts directly to a broad spectrum of users. In addition, it will provide the private sector an excellent opportunity to add value through customization and tailoring for their customers, especially in the area of transportation and road weather applications.

New road weather modeling and forecast tools should be developed to provide relevant, useful information to those who build, maintain, and use the nation's roadways. The road weather research program must support a range of model development and forecasting activities. For example, improvements are needed for models of weather in the microscale roadway environment. This advancement will require modifications to current numerical weather prediction techniques, which typically apply to much coarser spatial domains. Likewise, roadway transportation models need significant improvements to dynamically account for weather. Currently, most models of traffic flow ignore the impacts of weather. A long-term objective of the road weather research program should be to develop end-to-end models that seamlessly integrate real-time observations of weather and traffic, now-cast and forecast models for weather and traffic, and decision support systems for applying the model output to user needs. These end-to-end models will build upon the advances in the meteorological and transportation modeling efforts. The following sections describe this range of modeling efforts in more detail.

Modeling the Microscale Roadway Environment

Energy-balance models provide information useful to road managers on both tactical (< 6 hours) and strategic planning (> 6 hours) time scales (Crevier and Delage, 2001). These models can be strengthened by constraining them with assimilated weather and road condition data. Even if road weather models have imperfect parameterizations, accurate and well-placed road weather information systems will lead to improved forecasts. In the short term, maintaining an accurate, up-to-date database describing road and bridge properties (e.g., composition, structure, depth, roadbed characteristics) for the network of interest would enable more accurate estimates of the thermal characteristics and thus should result in improved forecast accuracy. Such databases could be evaluated using qualitative methods such as comparing recorded road properties to model predictions and the results of road thermal mapping. The relevance of chemical reactions involved in road treatment needs to be examined and if necessary included in the energy-balance models. On a time scale of 5 to 10 years, continued expansion of RWIS installations to areas identified as critical either because of known hazards or poor model performance would lead to forecast improvements.

Alternative semi-empirical approaches to determining the energy balance at the road surface, such as expert systems, also show promise, particularly in the context of improved RWIS, pavement sensors, and weather-forecast inputs. These will be particularly useful for vulnerable areas, such as bridges or ravines, where local conditions frequently create hazardous conditions but the local physics compromise the validity of the one-dimensional energy-balance approach.

Modeling of conditions that produce dangerous winds or drifting snow is an important consideration for many rural areas. Wind-tunnel (physical) modeling, large-eddy simulations, or empirical approaches could be used for this purpose, as well as for designing mitigation strategies (e.g., built or living structures for protection from drifting snow).

In the intermediate to longer term (10 to 15 years), improvement in road surface forecasts will require more accurate mesoscale weather forecasts and better representations of the local heat, moisture, and momentum exchanges. Such improvements will enhance abilities to simulate localized weather phenomena, such as fog development and nocturnal drainage flows. Improved modeling of both mesoscale fields and local energy balances under conditions of strong vertical gradients in the atmospheric boundary layer or significant temperature differences between the atmosphere and the surface is especially needed.

Many factors that affect the roadway environment (See Table 3-1) are represented in the newer roadway models through parameterizations. Following the development of such parameterization schemes, they should be tested and refined using further observations. Such testing should concentrate on individual factors (e.g., traffic intercepting infrared and visible radiation, how much heat is generated by vehicles), common but simpler combinations of factors (e.g., traffic on a dry road, traffic on a wet road with steady precipitation), or the final model product (e.g., road temperature). Measurement of the heat loss from the road to the atmosphere will be a challenge in the heterogeneous roadway environment, but Wojcik and Fitzjarrald (2001) have been able at least to make estimates within useable limits.

Recommendation: Improve empirical and numerical modeling techniques to account for both the roadway atmosphere interface and the surrounding environment.

> *The roadway environment consists of two components: (1) the road surface and (2) the weather and visibility conditions directly above the road. For the former the committee recommends that research be pursued to improve one-dimensional modeling techniques for the roadway-atmosphere interface, the region that extends a few meters above and below the road surface. Advances in this area would enable better forecasts of road surface temperature, leading to better forecasts of frost and black ice. For the latter, empirical and multidimensional modeling approaches should be developed to account for local terrain, land cover, surface composition, and other geographic features that affect the exposure of the roadway to infrared and solar radiation (direct and indirect), wind, and precipitation. Such models need to provide improved forecasts of road surface temperature and condition (e.g., dry, wet, frozen surfaces) as well as fog, wind, and other localized atmospheric phenomena.*

The information required for decision makers, like the available options, varies according to the time frame. Builders planning new roads in snow-prone areas will want to know where the combination of terrain and wind indicate a need for snow mitigation devices (snow fences). Bridge construction crews that want satisfactory conditions for the curing of concrete may want a projection of temperature over the first few days after the concrete is to be poured (Wojcik and Fitzjarrald, 2001). Trucking companies can use 1- to 3-day weather forecasts to plan cross country routes. Road managers deciding whether to put their snowplow crews on call would use the 18- to 24-hour projection of the timing and severity of a predicted snowstorm. Over shorter times the road manager would be interested in how the weather, snow mitigation strategies, traffic, and roadway properties affect roadway performance and safety.

In the tactical 0- to 6-hour time frame a mix of observations and road weather models is needed to provide the most useful results. In the absence of such models empirical relationships of road temperature or slickness at a given location to weather conditions can also be useful, especially for the shorter time frame changes. Indeed, simple measurements of road surface temperature, images of fog or snow on roadway Web cams, or reports of accidents can override model predictions on the less-than-2-hour time frame in decisions to keep a road open or closed. As lead times approach 6 hours, current data become less useful, and some models may not have "spun up." For such situations, ways to optimize the information available at these intermediate time scales must be developed. This likely will be the domain of the human forecaster, especially in cases of severe and hazardous weather, until the objective techniques become skillful within this "gap" period between observations and numerical weather predication forecasts.

Recommendation: Improve accuracy and resolution of road weather forecast products supporting both tactical and strategic decision making.

The road weather research program should support development of tailored road weather information products to support real-time operations and both near-term tactical (0 to 6 hours) and longer-term strategic (> 6 hours) decisions by road users and managers. More accurate weather information products should be developed on fine spatial and temporal scales appropriate to the roadway environment, in part by taking advantage of advances in nested and ensemble models. Producing road weather forecasts for the 2- to 6-hour timeframe presents the greatest research challenge because current data become less useful and some models may not have reached the point where they can provide useful output. As a result, this short-term forecast problem likely will remain a focus of the human forecaster and newly emerging statis-

tical tools for the near future. The newly developed National Digital Forecast Database from the National Weather Service provides forecasts across the United States on grid spacing no greater than 5 km and may provide a logical starting point for either user-developed applications or further enhancement by the private sector.

It is possible that current approaches to attaining finer spatial resolution information from numerical weather prediction models will not yield forecasts of sufficient accuracy for the roadway environment. These approaches include developing finer resolution versions of existing models, downscaling output from mesoscale models, and nesting methodologies. It may be necessary to pursue alternative methods for producing road weather forecasts. These might include statistical applications such as neural nets, empirical models, or probabilistic estimates derived from ensemble forecast systems. The leadership of the road weather research program must support research to determine the best strategy for maximizing predictive skill at minimum cost, this leadership will also have to determine whether a numerical weather prediction model development approach that focuses on downscaling of the conventional mesoscale modeling technology is the correct strategy.

Roadway Transportation Models That Account for Weather

The transportation community has long recognized the relationship between weather conditions and the operation of the transportation system; however, for many reasons the impacts of weather have generally not been incorporated into the tools used by transportation professionals to evaluate alternative transportation strategies.

Transportation planners and engineers use a variety of tools to plan for and operate the transportation system. Development of numerical traffic models began in the 1950s with simple car-following algorithms, followed by regional travel demand forecast models in the 1970s (macroscale), and evolved into complex models that simulate traffic flow and individual vehicle movement for major traffic corridors in a metropolitan area (mesoscale), along a given traffic *route*, such as a stretch of interstate highway (microscale), or flow through a given complex highway *interchange* (picoscale). Such models are used to identify the most effective physical improvements to make and to determine the optimum strategies for operating the system.

To date, these models have predominantly been used to make planning decisions. In planning the system, alternatives are studied for how best to accommodate demand on a portion of the transportation system. Such options as adding lanes, adding turn lanes, and designating lanes for special purposes (e.g., commuting) are analyzed. More recently traffic models have been used in a few places to operate the system, providing predictive traffic information for real-time traffic management. Such alternatives as differences in signal timing or ramp metering can be evaluated and immediately implemented. Strategies for best handling of incidents, such as accidents or flooding, can be analyzed with these models.

While these models consider many variables in their analysis, and while weather is known to be a significant variable, virtually all the tools assume the system is operating under acceptable, dry weather conditions. Considerable research has been conducted to identify the impact of factors such as lane width and shoulder width, but little research has been done to establish the impact of weather conditions on the operation of the transportation system. Significant research in this area is needed to conduct sensitivity studies over a plausible range of weather conditions. For example, efforts are needed to determine the relationship between rainfall intensity or duration and key traffic variables, such as roadway capacity, maximum speed, acceleration or deceleration rates, car-following parameters, lane-changing parameters, headways and startup times, and the range in driver speeds, all of which are important in evaluating transportation options.

Recommendation: Integrate available weather information into traffic planning and management models.

The committee recommends research to improve the way weather effects are included in traffic and emergency management models, which currently are used for offline design and planning of transportation systems. Better accounting for weather in these models will allow researchers to assess the benefits of improved weather monitoring and forecasts on transportation operations and enable practitioners to more efficiently manage the nation's roads.

The committee believes that over the next 15 years there is significant opportunity for the roadway community to couple atmospheric and traffic simulation models for real-time management of the transportation system. Such an initiative would not have been possible 10 years ago because numerical weather prediction and traffic simulation models were not sufficiently mature to be linked efficiently. That is not the case today. Indeed, traffic simulation models that can ingest real-time data—such as incident reports, traffic flow data, and work zone status reports—and even predict near-future traffic conditions are being developed. Two such models are the Dynamic Network Assignment-Simulation Model for Advanced Road Telematics model (DYNASMART-X) and the Dynamic Network Assignment for the Management of Information to Travelers model (DynaMIT). These models try to predict the behavior of vehicles by simulating thousands of driver choices every few seconds and predicting when those choices will converge to create congestion (Vatalaro, 2001). These models show great promise for supporting traffic management in developing real-time routing and rerouting scenarios for more efficient system management, but they do not ingest weather data.

Coupling atmospheric and traffic simulation models provides opportunities to address significant problems related to roadway safety, capacity, and efficiency. Some of the areas of opportunity are as follows.

- *Congestion Mitigation*: Coupled models for urban roads to detect and predict weather elements that contribute to congestion assimilate the dynamic data into the traffic simulation model and modify the traffic flow accordingly. The resultant output could trigger mitigation strategies, such as access control, ramp metering, speed management, weather-related incident detection, weather-related traffic signaling, and public advisories.
- *Emergency Evacuation Strategy Development*: Coupled models covering regional areas that contain urban areas, nuclear power plants or other facilities that have potentially hazardous releases, and hurricane-prone stretches of the U.S. coast could predict the movement and concentration of hazardous materials by local winds that feed into traffic simulation models that identify safe and unsafe evacuation corridors. These identified areas could trigger mitigation strategies, such as contraflow, access control, incident response and public advisories to facilitate the movement of people away from the danger in the most efficient manner. Use of the system for hurricane evacuation would identify similar hazardous areas to be avoided and initiate similar mitigation strategies, particularly contraflow and access control.

Although both model development paths are fairly mature, there is still much research that is required to tap the opportunities outlined above. Work needs to be done to determine how to best use weather information in traffic simulation models to allow them to further optimize traffic flow for a range of weather conditions in a robust and stable manner.

A phased approach to development is likely the best path, beginning with more straightforward problems, including simple prototypes, and then migrating to more complex problems; for example, determining the effect of light, moderate, and heavy precipitation on traffic flow along certain corridors could be a beginning point. More complex problems, such as drifting snow on the roadway and crosswind effects on high-profile vehicles, would be addressed at a later stage. Given the current strong focus on homeland security, it might be appropriate to begin the research with coupled systems that take a simple, first-guess estimate of a toxic plume's location and movement and determine how this information can be effectively used in a traffic simulation model. Predictions of plumes that are highly complex in their movement to local winds and atmospheric stabilities could be addressed later as confidence in the coupled system increases.

Recommendation: Use real-time weather information in the operation of the transportation system.

> *The road weather research program should support development of operational modeling systems for roadway management that include real-time weather information. The use of real-time weather information to support the operation of the transportation system is in its infancy but should be encouraged; for example, the road maintenance community has made advances in incorporating real-time weather information in decision support systems for winter operations. Real-time road weather information would also be useful for warning drivers of dangerous weather conditions and for optimizing commercial vehicle operations.*

End-to-End Models

Decision support systems are the largest category of weather-related needs for the transportation system manager. For decades weather information has been assimilated by these managers using a method that some refer to as "swivel chair integration." That is, the decision maker considers each separate category of weather-related information independently and then tries to integrate the accumulated information manually to make a critical decision. Many managers have learned to do this quite well, but their decision making process will almost always be different than their colleagues; this means that inherent inconsistencies exist across the team of decision makers in any given facility.

If one adds to this a process that is very inefficient by its nature, there is no wonder that most facilities have come up with a consensus process that is conservative for the group. In other words each decision associated with a weather scenario is not optimized in a benefit-cost sense. Winter maintenance rules of practice, for example, tend to be too conservative, usually resulting in a higher than required margin of safety and the application of more chemicals to the roadway. This practice results in higher costs and negative effects on the environment.

The roadway management decision process is further complicated by the fact that weather is only one of several decision criteria that need to be assessed. A reliable, consistent decision maker must integrate all these factors to be successful. Modern decision support systems can integrate all this information consistently and reliably, but use of such systems is rare in today's roadway management systems.

The need for decision support applications is widespread, even in the traffic management community. Some of the needs are as follows.

- Integration of weather-based decision support systems with traffic simulation numerical models is required to diagnose and predict traffic flow in urban areas for short periods into the future. The primary weather-related information needed is pavement condition (e.g., wet, dry, icy, snow-covered), precipitation type and rate, wind speed and direction, road friction measurements, and obstructions to the driver's vision (e.g., sun glare, fog, smoke, dust, or blowing snow). This decision support information is needed to advise motorists of predicted and prevailing conditions, to alter the state of roadway devices to control traffic flow and regulate roadway capacity, and to supply resources to roads to mitigate weather impacts.
- Decision support technology is needed to trigger warnings of dangerous conditions along the roadway resulting from weather events. Some examples are reduced driver visibility ahead, cross-winds exceeding 50 mph ahead that can tip over high-profile vehicles, severe thunderstorms (hail, heavy rain, high winds), tornadoes, icy road conditions ahead, flooding on the roadway, hurricane conditions, movement of wildfires near the roadway, and the movement of chemical, biological, or nuclear hazardous material along the roadway (OFCM, 2002a).
- Decision support technology is needed to optimize commercial vehicle operations. This would enable truck routing to exploit weather information for just-in-time delivery and to minimize fuel consumption and allow the safest routing for hazardous cargo shipments based on road condition and winds relative to vulnerable assets and temperature.

An end-to-end road-modeling system has enormous potential to enable road managers and emergency responders to develop better responses to everything from developing strategies for safer driving in fog and responding to extreme conditions, such as evacuating the coastline in anticipation of hurricane landfall. The required modeling system would have to incorporate changing weather conditions, preferably with high horizontal and temporal resolution to account for local variability; a model of road conditions, especially road temperature and slipperiness; a traffic simulation model that includes the reaction of drivers to such conditions as precipitation, wind, and traffic; and traffic mitigation procedures, such as timing traffic lights, switching direction of highway lanes, and dynamic message signs.

Recommendation: Develop end-to-end models that assess and predict weather impacts on roadway conditions and operations.

The road weather research program should support a long-term, interdisciplinary effort to develop end-to-end road-modeling systems. The required modeling system would have to incorporate (1) current weather and pavement conditions; (2) "future" or forecast weather conditions; (3) a model of road conditions, especially road temperature and traction; and (4) a traffic simulation model that includes the reaction of drivers to such conditions as precipitation, high winds, low visibility, slick pavement, and congestion mitigation strategies (e.g., timing traffic lights, commuter highways, weather-controlled dynamic speed limit signs). Used in an offline planning mode, such a system has enormous potential to enable road managers and emergency responders to develop better proactive responses to extreme weather conditions. As understanding, modeling sophistication, and computer capabilities allow, such models ultimately could be run in real time to assist in the routine management of the transportation system, thereby enhancing safety, capacity, and traffic flow.

COMMUNICATION SUPPORT FOR ROAD WEATHER INFORMATION USERS

The road weather research program should support the development of multiple mechanisms for communicating road weather information to a range of users in a manner that supports more informed decision making. Communication is important for making the research results useful to drivers and those who build, maintain, and operate the roadways. Development of improved mechanisms for communicating weather information to roadway users is already under way: traffic and weather information is often relayed together in radio and television broadcasts; many states have implemented 511, a single nationwide phone number for traffic and road condition information; and roadway managers are making greater use of dynamic message signs. The Maintenance Decision Support System pilot project has developed a prototype tool with which winter maintenance operators can obtain highly targeted road weather observations and forecasts, as well as assistance in choosing the best treatment options.

The committee sees a road weather research program in which two-way communication allows researchers to learn from stakeholders about their needs and thus build better decision support tools for them. In addition to setting up effective mechanisms for communication, the road weather research program should support research on ways to improve communication. Two areas on which this research should focus are improving the ability to convey levels of confidence associated with road weather information and taking advantage of newly available communication tools.

Conveying the Confidence Level

Testimony to the committee by both users and providers of advanced weather technology indicated a strong need for conveying not only timely, accurate, and reliable diagnostic and prognostic information to decision makers, but also the degree of uncertainty in this information. Users have historically shied away from probability metrics associated with forecasts, in some cases because the need is misunderstood and the metrics are considered merely a way for prognosticators to dodge their real need: a perfect weather forecast out to a week or so. Even though the user might have understood the need for the uncertainty metric, intelligent use of the information was another problem; very few users had systems that could ingest this information and analyze its value to the decision maker. Now, advanced weather decision support systems that assimilate many diverse sources of weather and operational databases have uncertainty metrics as a by-product, and in many ways depend on the metrics for their value to the user, for example, cost analyses can be done effectively by ingesting probabilistic weather data along with information about available resources and operational status.

In addition to developing robust methods for estimating the degree of confidence in road weather information, it is important to devise effective methods of conveying uncertainty information to users. Communicating uncertainty was the subject of a recent National Research Council workshop during which lessons learned from several weather and climate case studies were discussed (NRC, 2003a). The workshop participants found that communicating uncertainty is a critical part of decision making for users of weather information, and it is important to communicate both why information is uncertain and why that uncertainty matters. The workshop report stresses that multiple measures of uncertainty, both quantitative and qualitative, should be communicated, along with a context to tie the information to past experiences and to make the uncertainty information more tangible.

Despite advances, much research is needed on conveying uncertainty in road weather information products and improving easily understood metrics to facilitate user decision making. New ways are needed to include a more complete representation of the sources of uncertainty and to communicate the context of the information. The format of information (e.g., text, graphics, numbers) may serve to modify its interpretation even when the underlying information is unchanged. Simply repeating the same information tends to convey that it is more credible. Recent memory of similar events may color interpretations, whereas prior education about how to interpret information can improve decision making. Even though the NWS has developed products to provide uncertainty information to address some user needs (e.g., the probability of measurable precipitation or the seasonal and monthly outlooks), much work needs to be done with users to address the full expanse of their needs and determine the best ways to convey uncertainty to them. This is particularly true with the newly implemented National Digital Forecast Database, which presents very high resolution (5 km grid spacing) forecasts through seven days without the benefit of confidence or uncertainty information.

Recommendation: Develop methods for estimating and conveying the degree of confidence in road weather information.

> *The road weather research program should support the development of means for estimating the inherent uncertainty in road weather information. Advanced weather decision support systems that assimilate many diverse sources of information need to estimate uncertainty and provide easily understood metrics to aid users in making optimal decisions. New methods are needed to enable more effective communication of uncertainty to a wide variety of users.*

New Communication Tools

Systematic collection and dissemination of road weather data are critical to the success of the road weather research program and require robust communication technology. For example, collection of real-time data invariably involves some form of data logger at the measurement site, telecommunication means (e.g., telephone line, fiber line, microwave relay), and a reception facility that moves the inflowing data stream to archival servers. These communication means are often already in place in roadway right-of-ways, either for commercial communications providers, law enforcement, or state departments of highways themselves.

A number of new wireless technologies may enable communication with motorists when and where they need it. The technologies that hold the most promise are satellite radio, cellular, and wireless local area networks. When considering how each of these technologies may be used for communicating road weather information with drivers, it is necessary to consider how much information can be transmitted and whether it is possible for two-way communication, how far the signal can be transmitted, and how ubiquitous the technology is. Also weighing into the decision are human factors, such as familiarity, ease of use, and cost. Each of these technologies is capable of transmitting different bandwidths, which affects the type of information that can be sent.

Satellite-based Digital Audio Radio Service, which provides radio coverage across the continental United States, has been implemented recently. With a satellite-capable radio receiver and a subscription to a satellite radio service, consumers at home or in their automobiles are able to receive detailed graphic weather and traffic reports that are customized for their location. Although this technology currently is available only to emergency response vehicles, it may become commercially available soon. Several automobile manufacturers are beginning to offer satellite radio as an available factory option. Satellite radio is an improvement over regular radio because it has better coverage and bandwidth, while allowing geolocated transmissions. Nevertheless, regular radio remains critical for transmitting weather information to drivers because the costs to consumers are minimal.

Cellular communication provides many features that match the needs for communicating weather information to drivers. In most parts of the country, cellular signals have a range of miles (see Figure 4-6) and a relatively ubiquitous infrastructure in place, with many of the cellular towers installed along interstate highways. The technology is flexible and the focus

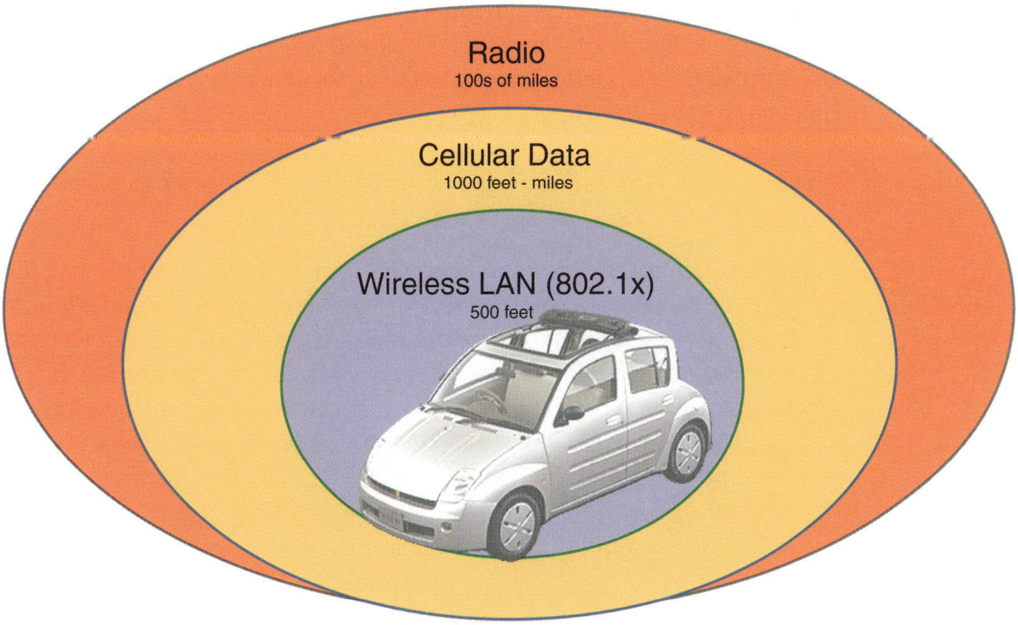

FIGURE 4-6 Schematic illustrating the spatial range over which different communication technologies extend. SOURCE: Ian Ferrell, Microsoft Corporation.

of much innovation, making it likely that applications for the road weather problem might be developed. In particular, it is possible to send individualized broadcasts, for example with information about weather on an adjacent stretch of road, from each cell tower. Again, human factors, such as familiarity, ease of use, and cost will play a role in the feasibility and success of such applications.

Wireless local area networks have a much larger bandwidth than cellular and recently were tested successfully at highway speeds of over 200 mph (http://wireless.itworld.com/4260/031020necwifi /page_1.html). Implementation of this technology may occur soon, but until then wireless local area networks are well suited for in-car communications with other devices and for communication between the vehicle and physical points, such as a home, gas station, or rest stop. Very detailed information about local weather and road conditions, including animated forecasts or other large-bandwidth data, could be transmitted to a vehicle located at a rest stop. As with cellular, wireless local area networks support two-way communication and are the focus of much technological innovation; however, little infrastructure has been developed to support wireless local area networks communication with vehicles, so coverage is very poor at this time.

Recommendation: Develop new means to effectively communicate road weather information to a wide range of users.

The road weather community should take advantage of new and upcoming advances in technology that will enhance communication to users. Along with expanded observational capabilities, advanced communications technology must be deployed to enable instruments to transmit data back to central processing centers and to relay processed weather information back to drivers and other users. Advances in wireless communication hold much promise for providing location-specific weather information to mobile users, but the challenge of determining how best to use these communication technologies remains.

INTEGRATING WITH THE EMERGING INFOSTRUCTURE

A final but nonetheless critical emphasis of the road weather research program should be to develop a telecommunication infrastructure that takes advantage of new technologies to effectively convey road weather information to end users. Previous recommendations in this report have included the establishment of a nationwide real-time road weather observation system. Such a national integrated network of fixed and mobile stations would generate an enormous amount of valuable information about the weather affecting roads and travelers and the resultant condition of the road network. All of these data would need to be carefully and uniformly quality controlled to common national standards to ensure consistent and appropriate use by multiple agencies (also a prior recommendation). The meteorological community, both public and private, would make use of this rich dataset to refine numerical weather predictions of the atmosphere and determine future impacts on the roadway environment (Smith, 2001). Weather impacts permeate the entire transportation system and must be factored into all manner of road transportation operations and systems. Clearly, there is a substantial telecommunications requirement to first collect road weather and condition data from multiple fixed and mobile platforms and then move road information and finished products to transportation managers and road users.

A national telecommunications and information technology infrastructure or "infostructure" is required to move all this information. This infostructure must be adapted to respond to the specific needs of the road transportation sector. The National ITS Architecture, discussed in Chapter 3, provides the overall blueprint for this infostructure. It starts out by identifying the basic elements of road transportation and then groups them into four categories or systems: travelers, vehicles, field, and centers. Within each of these systems, specific elements or subsystems with unique needs are defined; for example, the vehicles group includes maintenance and construction, transit, commercial, emergency, and regular passenger vehicles. Subsystems are defined, based on extensive user interviews, for the other main systems: centers, field, and travelers.

The National ITS Architecture then goes on to define basic telecommunications links between all of these systems. These are grouped into four classes: (1) from the centers to the field, regular wireline (fixed-point to fixed-point) communications suffice and are specified; (2) between the field and vehicles, Dedicated Short Range Communications provides the solution from fixed points to moving vehicles; (3) vehicle-to-vehicle communications; and (4) wide-area wireless communications.

The very serious safety and efficiency impacts that weather exerts on road transportation are recognized and have been included throughout the Architecture. Each of the fundamental activities leading to better management is covered: data acquisition, processing, dissemination, and integration into end-to-end applications. In responding to the needs of road transportation the meteorological community must take full advantage of the infostructure as specified by the Architecture. Road weather and condition data are collected from multiple fixed sites and from vehicle fleets hosting road and weather sensors. Links are specified to move the data to primary operators of the sensing equipment, the road maintenance and construction community. The architecture then provides for the data to be moved to other agencies for processing into useful information. This includes quality assuring the data, analyzing them, generating road weather forecasts from them, and finally packaging and returning all this information to all manner of road transportation users. The applications range from integrating time and location information into future in-vehicle audio and video display systems to ensure rapid and easy assimilation by motorists while driving to integrating high-resolution road weather modeling outputs directly into end-to-end solutions for road maintenance and traffic and emergency management.

The transportation community has taken the first and most important step of recognizing the impacts of weather and the need to factor weather information into their operations. The National ITS Architecture incorporates design elements for road weather telecommunications. The weather community needs to work through the Architecture with transportation professionals to respond to road transportation's specific needs. This process is already engaged. Weather and transportation professionals worked together on ITS America's Weather Information Applications Task Force and the FHWA's own Maintenance and Construction Operations (MCO) User Forums to develop the MCO User Service. Other activities are currently being pursued in the standards area, another main application area for the Architecture. Such standards as the National Transportation Communication for ITS Protocol Object Definitions for Environmental Sensor Stations have benefited from the participation of weather and transportation professionals. That standards effort is nearing maturity, but others are at an early stage and will need the participation of both communities in order to succeed. Numerous other future applications will require both weather and transportation expertise.

The transportation community is developing, with the National ITS Architecture as the fundamental design, a robust national infostructure adapted to the specific needs of road transportation. That infostructure presents an enormous opportunity for weather professionals to work with the transportation community to prepare and effectively deliver uniquely adapted road weather products. It is therefore crucial for the meteorological community to develop road weather professionals and engage the transportation community in the development of weather solutions for road transportation.

Recommendation: Develop a robust national roadway infostructure.

Major components of a roadway infostructure (a network of data collection and dissemination necessary to support real-time management and operation of the roadway transportation system) are being developed. The committee recommends that the road weather research program proactively participate in this effort to ensure that the road infostructure of the future incorporates a sophisticated network of road weather observations, including sensors embedded in the pavement, weather stations adjacent to the roadway, water level sensors near flood-prone routes, remote observations from satellite platforms, and instruments on vehicles themselves.

5
Implementing Improved Road Weather Research and Management Programs

Coordinating the research and services described in Chapter 4 will entail a complex management structure that engages the many entities conducting road weather research and related services and products. Up to the present, research relevant to road weather problems has been conducted largely independently in the meteorological and surface transportation communities. Many road weather services and products have been produced by the private sector and implemented by individual states. To attain the vision for the road weather system of the future, laid out in Chapter 2 of this report, the committee believes that a focused, coordinated research program is needed. Many of the research needs and opportunities that this program should address are presented in Chapter 4. In this chapter existing research programs are described, followed by a discussion of why they are not able to meet the future needs for road weather research. Options for an organizational, management, and funding strategy to meet future needs are then presented. Finally, the need to develop robust technology transfer and education is discussed.

CURRENT ROAD WEATHER RESEARCH EFFORTS

Highway Research Programs

Highway research is coordinated by the Federal Highway Administration (FHWA), the states, the National Cooperative Highway Research Program, academia, and the private sector. Federal funding, derived largely from dedicated revenues in the highway trust fund, supports FHWA research, more than half of state highway research, and the National Cooperative Highway Research Program. These programs coordinate with each other informally, and in some cases through formal partnerships, though more commonly through professional relationships and collaborations

TABLE 5-1 Funding for Surface Transportation Research

Transportation Research Program	FY 2001 ($ in millions)
Federal Highway Administration	
Surface transportation research	86.1
Technology deployment	39.6
Training and education	15.8
Intelligent transportation systems	42.5
University transportation centers	23.9
State Highway	
National Cooperative Highway Research Program	30.6
Other federal funds	185
State funds	153
Private Sector	75–150
Total[a]	621–696

[a]Total federal funds spent on surface transportation in FY 2001 were $393 million.
SOURCE: TRB (2001b).

among decision makers, researchers, program managers, and state highway personnel. See Table 5-1 for a summary of research funding for surface transportation.

In FY 2001 the FHWA spent nearly $184 million on surface transportation research, technology deployment, training and education, and intelligent transportation system research and development (TRB, 2001b). About $2.5 million of this funding is devoted to road weather research and development (see Table 5-2). The FHWA also supports the Turner-Fairbank Highway Research Center, the Highway Safety Information System database,[1] and the Intelligent Transportation Systems Joint Program Office. The FHWA conducts some research itself and contracts for the remainder with private firms, universities, and research institutes. The administration typically selects its research contractors through merit-based competition. In recent years, however, Congress has designated more research projects and research performers. Congressional designations under the Transportation Equity Act for the 21st Century for 1999, 2000, and 2001 amounted to 44

[1]The Highway Safety Information System database includes crash, roadway inventory, and traffic-flow data from eight states and is used to study safety issues in roadway design, maintenance, and safety treatments.

TABLE 5-2 Funding for Federal Meteorological Research (in millions of dollars)

	FY 2001	FY 2002	FY 2003
Department of Agriculture	15.5	15.5	32.9
Department of Commerce	102.7	109.2	102.1
Department of Defense	97.8	72.3	54.0
Department of Transportation	27.8	26.4	27.5
Aviation	*25.1*	*24.0*	*25.0*
FHWA	*2.7*	*2.4*	*2.5*
Environmental Protection Agency	6.4	6.4	7.5
National Aeronautics and Space Administration	165.7	154.3	154.5
Aviation	*35.3*	*35.3*	*55.7*
National Science Foundation	188.9	202.0	218.9
Total[a]	604.8	586.1	597.4
Aviation	*60.4*	*59.3*	*80.7*
Other Transportation	*2.7*	*2.4*	*2.5*

[a]Although the Department of Energy does conduct meteorological research, they do not specifically report the amount in their annual budget, thus their portion is not included in the total.

SOURCE: OFCM Federal Plans for Meteorological Services and Supporting Research and the National Science Foundation Summary of Budget Request to Congress from FY02 (OFCM, 2001; NSF, 2002), FY03 (OFCM, 2002a; NSF, 2003), and FY04 (OFCM, 2003; NSF, 2004).

percent, 42 percent, and 51 percent of the FHWA's research and technology spending, respectively (TRB, 2001b). According to a recent report of the Transportation Research Board, the FHWA's research program has produced "incremental improvements leading to lower construction and maintenance costs, better system performance, added highway capacity, reduced highway fatalities and injuries, reduced adverse environmental impacts, and a variety of user benefits" (TRB, 2001b).

The Department of Transportation's Research and Special Programs Administration administers the University Transportation Centers program, which was initiated under the Surface Transportation and Uniform Relocation Assistance Act of 1987. The act authorized the establishment and operation of transportation centers, designated through a competitive process, in each of the 10 federal regions. The program was reauthorized by

Intermodal Surface Transportation Efficiency Act and again by the Transportation Equity Act for the 21st Century, which provided up to $158.8 million for grants to establish and operate as many as 33 centers throughout the United States in FY 1998 to 2003. University programs must match the federal funds they receive, usually on a one-to-one basis, with federal, state, or other funds. Ten of these centers, designated as regional centers, were selected competitively in 1999. The other 23 centers are located at universities specified in the Transportation Equity Act for the 21st Century.

Each state highway agency conducts applied research to address technical questions about planning, design, construction, rehabilitation, maintenance, and environmental conditions of highways in the state. Projects are typically expected to produce rapid results that can be immediately applied to specific problems within the state. A 1999 survey by the American Association of State Highway and Transportation Officials (AASHTO) found that approximately $322 million was spent by states on research that year, $144 million of which were federal funds provided through the State Planning and Research Program and $32 million were from other federal funds made available for state research.

Many states also participate in pooled-fund projects, which leverage the limited funds available in individual states to tackle problems of mutual interest. The largest pooled-fund project is the National Cooperative Highway Research Program, a voluntary program established in 1962. The states contribute 5.5 percent of their State Planning and Research Program funds to the National Cooperative Highway Research Program, a total of $30.6 million in FY 2001. General topic areas for projects are selected by AASHTO's Standing Committee on Research with assistance from its Research Advisory Committee and approval by its board of directors. The Transportation Research Board convenes expert panels to select the researchers for each topic.

Research is also conducted by a number of private sector companies and associations. Their research agendas are largely problem oriented and profit driven. Funding for private sector research in 2001 was estimated to be $75 to $150 million (TRB, 2001b).

The highway research community has conducted a number of special highway research initiatives designed for problems that are particularly challenging or are ready for significant advances. The most recent example was the Strategic Highway Research Program, a five-year, $150 million program authorized by Congress in 1987. The program was recommended by the Transportation Research Board to improve the performance of highway materials and highway construction and maintenance practices (TRB,

1984). The program yielded substantial advances in winter maintenance practices, particularly by identifying the value of RWIS proactive anti-icing and evaluating the effectiveness of several new chemicals to control snow and ice.

A recent Transportation Research Board report requested by Congress identified several topics for a follow-on to the Strategic Highway Research Program, named the Future Strategic Highway Research Program. The report recommended a research program aimed at developing a systematic approach to highway renewal, preventing or reducing the severity of highway crashes, providing highway users with reliable travel times, and systematically integrating environmental, economic, and community requirements into the analysis, planning, and design of new highway capacity. The proposed efforts are currently being considered for inclusion in the reauthorization of the Transportation Equity Act.

Some highway research is also conducted by the federal government through the Federal Motor Carrier Safety Administration, the National Highway Traffic Safety Administration, the Federal Transit Administration, and the U.S. Department of Transportation's Research and Special Programs Administration. Intermodal research projects that address highway issues are conducted by the U.S. Department of Transportation's Office of Intermodalism and in some cases by other agencies such as the Federal Aviation Administration and the Federal Transit Administration.

Meteorological Research Programs

Similar to the case for transportation, meteorological research is conducted by many different entities, although in the case of meteorology most research is funded by the federal government. Reflecting the many effects that weather has on human activities, seven federal agencies support meteorological research: the Departments of Agriculture, Commerce, Defense, and Transportation; the Environmental Protection Agency; the National Aeronautics and Space Administration; and the National Science Foundation. The Office of the Federal Coordinator for Meteorology coordinates research and operational activities conducted by individual agencies.

The total federal expenditures on meteorological research in FY 2003 were $597 million (see Table 5-2), with the largest portions contributed by the National Science Foundation (37 percent in FY 2003) and the National Aeronautics and Space Administration (26 percent in FY 2003). The Department of Commerce, which houses the National Oceanic and Atmospheric Administration, National Weather Service, National Environmental

Satellite Data Information Service, National Ocean Service, and the Office of Atmospheric Research, has the largest overall expenditures on meteorology when operational activities are considered ($1.529 billion in FY 2003). Only 7 percent of the Commerce meteorological budget, or $102.1 million, is devoted to research. Much of the federal funding is used to support universities and nonprofit research institutes, such as the National Center for Atmospheric Research, which conducts the majority of meteorological research.

States and the private sector also support some meteorological research, albeit less than the federal government. Most states have a state climatologist and also contribute funding to regional climate centers and to state agencies with meteorological requirements. Private sector meteorology is very operationally oriented and invests a significant but unknown amount of money in developing new meteorological products and services for the media and industry, and in providing consulting services to lawyers, engineers, and architects. About 400 commercial weather companies and independent contractors operate in the United States, with revenues of about $500 million (NRC, 2003b). Market incentives encourage private companies to generate innovative products and ways of presenting weather information. These products are usually developed to address specific customer requirements.

The U.S. Weather Research Program is a partnership among several federal agencies, the American Meteorological Society, academia, and the private sector intended to coordinate efforts among these groups to accelerate forecast improvements of high-impact weather and facilitate full use of advanced weather information. The program aims to reduce the effects of weather-induced disasters, minimize the costs of routinely disruptive weather, take advantage of improved weather information, and assist the military. The current federal agencies participating in the program include the National Oceanic and Atmospheric Administration as the lead agency, the National Science Foundation, National Aeronautics and Space Administration, Department of Energy, Department of Transportation, Department of Agriculture, and the navy (USWRP, 2003). The program is looking to broaden its membership, particularly to include agencies to which improved meteorological information could be beneficial (OFCM, 2002a).

CHALLENGES FACING ROAD WEATHER RESEARCH

Despite the significant funding devoted separately to both surface transportation and meteorological research, an important finding of this committee's investigation is that road weather research is greatly underfunded in both the transportation and meteorological sectors, in part because there is no permanent funding. As a comparison, approximately a combined $60 million a year has been devoted to aviation weather research for the past decade by the Department of Transportation and the National Aeronautics and Space Administration (see Table 5-2) resulting in a substantial decrease in weather-related accidents. It is reasonable to expect that a comparable investment in road weather could yield a significant reduction in roadway accidents occurring in adverse weather conditions.

In the preceding section various funding programs in both the surface transportation and meteorological research arenas have been described. Each community can be characterized as supporting a large number of worthy investigations competing for limited dollars. In general this is an environment that makes it even more difficult to develop an effective and lasting interdisciplinary research program. Road weather research proposals compete in a field of a large number of topics that more clearly fall within established transportation or meteorology funding programs. Investigators are often reluctant to fully embrace interdisciplinary programs, given the tendency of funding to be of limited longevity and, in some cases, the merit of the research not being fully recognized within one's discipline.

Addressing research and operational needs of the road weather problem requires an effective engagement of a variety of entities spanning the meteorological and highway transportation endeavor. Research will need to be conducted by experts in government, academia, and the private sector who understand the roadway environment, how weather affects it, what meteorological information will help improve surface transportation, and how to deliver such information to users on and off the roadway. Researchers will need to interact with many funding mechanisms and stakeholders from both the transportation and the meteorological communities. Critical governmental stakeholders include federal and state highway agencies, federal agencies that produce or use weather services, emergency managers, and agencies that represent other modes of transportation on the federal, state, and local levels. The private sector plays an important role in developing new instruments and technologies, in road construction and maintenance contracted out by government agencies, in providing highly targeted meteorological products to specific users, and in using information to improve operations (e.g., just-in-time delivery by freight haulers). Adding further complexity is the necessary extension of road weather research topics from purely operational arenas, with imposed real-time requirements, to unresolved scientific challenges requiring ongoing fundamental investigations. Effective technology transfer must also be established to avoid results languishing in the "valley of death"; that is, the transition period between research and operational implementation (NRC, 2000).

If road weather research is to advance, professionals in the transportation and atmospheric science communities who have not worked closely in the past and who often do not have a shared understanding of the issues must coordinate their efforts. There are often significant differences in the ways various communities communicate and in the ways they address scientific and engineering challenges. Interaction between researchers with different areas of expertise must be fostered by creating opportunities for meaningful collaboration and by articulating clear objectives of the road weather research enterprise that are understood and accepted by both communities.

One approach to fostering collaborations between experts is the use of short-term, targeted funding programs to address specific challenges. These efforts may be effective in advancing understanding of the specific objective but may not realize the full potential benefit of working across disciplines because researchers may seek funding to work on their part of the problem rather than working together on the problem. Furthermore, investigators generally retreat to their own discipline as soon as the programmed research is completed. In the road weather research area there have been a number of successful funding programs of this type, including the Strategic Highway Research Program of the early 1990s and the Maintenance Decision Support System in more recent years. Although these early starts are promising, this committee believes that road weather research would benefit from a more permanent funding structure and the development of a lasting alliance of the benefiting communities. Attaining this goal involves the growth of an interdisciplinary community with a common vision, a sharing of ideas and methods, and long-term involvement of institutions and investigators. This second outcome is more desirable although more difficult to attain. Creating research centers specifically to bring together the range of expertise needed to address the road weather problem is one effective way to build this type of interdisciplinary community.

In addition to creating opportunities for meaningful collaboration, new interdisciplinary research will benefit from a clear articulation of the objectives, priorities, and path to implementation. On this topic, lessons can be learned from the aviation weather research community, as it has developed what is now a relatively stable and effective program. The Task Force on Aviation Weather Forecasting was particularly important in defining an effective path (NRC, 1994) and the Federal Aviation Administration then took the lead in fully delineating the goals for a new national aviation weather program. Nationwide leadership was critical in focusing and coordinating the efforts of researchers at many different institutions. The road weather community is just beginning to develop such a focused research plan, largely under the leadership of the FHWA.

PROPOSED ROAD WEATHER RESEARCH PROGRAM

It is clear to the committee that road weather is a scientific and technical challenge that now can be met with appropriate policy, implementation of available technology based on current knowledge, and research that addresses the many areas where knowledge is lacking. Over the course of the next 15 years a focused road weather research program could deliver improved road weather understanding and technology to the nation, saving lives and money while increasing the efficiency of road transportation. Such a program, with clearly defined goals, could oversee a suite of research activities as well as efforts to foster the implementation of an operational road weather capability. The recommendations that follow in this chapter, combined with the more technical recommendations in Chapter 4, are designed to help shape and guide such a focused program.

The committee's recommended framework for a focused road weather research program is based on careful review of the current scientific and technological opportunities and constraints as laid out in Chapters 3 and 4 of this report. Also considered was the institutional environment that such a program would need in order to thrive, including interaction with multiple funding mechanisms, researchers from many disciplines, a variety of stakeholders, and practitioners in the public and private sectors. In the committee's view the recommended program elements could provide an effective structure for addressing these multiple concerns. The committee expects that program leadership would conduct careful cost-benefit and other policy analyses to determine an optimum strategy for implementing the program.

Recommendation: Establish a focused, coordinated national road weather research program.

The committee finds that there are substantial research questions and opportunities in road weather that warrant a long-term national commitment and therefore recommends the establishment of a focused, coordinated national road weather research program. Sufficient knowl-

edge and experience exist today to initiate such a program; however, some aspects of the program exist only in concept and thus will require additional research and experience before they can be completely defined and implemented. A road weather research program is timely in that it can take advantage of investments being made in weather and in transportation research and infrastructure. An incremental investment in integrating these efforts will reap substantial benefits by producing a national road weather information system as part of the nation's emerging infostructure.

The goals of the road weather research program should be to

- *maximize the use of available road weather information and technologies;*
- *expand road weather research and development to enhance roadway safety, capacity, and efficiency while minimizing environmental impacts; and*
- *effectively implement new scientific and technological advances.*

The committee believes that the goals of this road weather research program can best be met through a nationally led program that supports regional research centers, national demonstration corridors, and a nationwide solicitation to support individual investigator-led research projects. Establishing and sustaining such a program long enough to achieve its goals will require dedicated funding sources over at least 15 years.

The proposed program would initially seek to take advantage of current capabilities, both in separate research endeavors that have not yet integrated their efforts and in existing observational, laboratory, and computational infrastructure. As the program matures it would also support new research and development that would build on the early accomplishments. Such a focused program would foster the development of a more cohesive road weather research community, both by bringing together researchers from different disciplines and by creating opportunities to educate the next generation of road weather researchers.

At all stages research supported by the program would be closely integrated with the development of operational capabilities and road weather information products for specific stakeholders. It will be critical for the program to foster interaction between the research community and users of the research results and to integrate user needs into all stages of research and development. Indeed, the private sector can serve as an important conduit for transferring research results to operations.

Establish Regional Research Centers

The committee recommends the establishment of regional research centers to develop new technologies, foster technology implementation on regional roadways, and facilitate interaction between federal, state, and local governments, the private sector, and academia. Such centers are an attractive alternative to individual state programs because many states in a particular region share the same road weather challenges. These regional centers should be interdisciplinary, incorporating weather and transportation researchers as well as relevant practitioners in the public and private sectors. They should be funded competitively with due consideration given to reflect different weather conditions and transportation problems common to their regions.

Multiple regional research centers are preferable to a single national center because different regions of the country face quite different weather conditions. Regional centers enable much more direct interaction with stakeholders who ultimately use research results. In addition to addressing the needs of their regions, individual centers should specialize to avoid redundant research on the road weather challenges that are multiregional or even national in scale; for example, one center could specialize in fog as a visibility impediment (a national problem), while another could specialize in evacuation strategies (a problem affecting hurricane-prone regions). The selected regions may or may not coincide with federal transportation regions. Opportunities to coordinate with existing regional centers and other research initiatives at national laboratories, universities, and state departments of transportation should be considered when selecting regions to house these centers.

The committee believes that it will be necessary to establish several regional centers to adequately cover the breadth of road weather problems and provide sufficient interaction with stakeholders. The road weather research program should provide base funding that is dedicated to establishing and operating these research centers. This funding should support efforts to assemble a critical mass of researchers, acquire space and other infrastructure, and support stakeholder interaction. To ensure that these centers are doing the best possible research but also have enough stability to pursue longer-term research agendas, the committee recommends that support for the centers be awarded competitively with five-year contracts.

Because regional research centers may require substantial investment in infrastructure before research can be conducted, the center selection should consider collocating with and leveraging resources, such as computing centers and strong university research programs. An alternative to regional centers that avoids some of these initial costs is the concept of "technology development teams," which typically include researchers selected competitively from various research centers; the selection process can, however, sometimes favor larger research efforts over individual researchers. This concept has been used successfully by the Federal Aviation Administration's Aviation Weather Research Program to provide efficient and flexible operationally relevant research results. Because surface transportation has a broader, more distributed stakeholder community, it is not clear how effective technology development teams would be in addressing road weather. If the technology development team approach is chosen, strong and effective linkages to the private sector should be built into the team structure.

Establish National Demonstration Corridors

As a means both to demonstrate the effectiveness of road weather improvements and to facilitate nationwide implementation of research results, the committee recommends the establishment of national demonstration corridors along two U.S. interstate highways. Special programs should be established for the federal government to partner with states and their private sector entities to showcase road weather products and services. Many of the recommendations in this report for new technology should be implemented first along these corridors. A major objective for these corridors is for adjacent states to work together to provide a seamless stream of road weather information to users.

The corridors will provide opportunities for researchers to implement and test new technologies and tools designed to address weather problems. Because they would be specially designated, these demonstration corridors would be a highly visible and easily identifiable way to showcase the results of the research. Another advantage of these test corridors would be the ability to install a permanent enhanced road and weather observation network. Such a network would allow multiple projects to leverage this asset, and each project would only be required to install a reduced set of temporary observing platforms. Establishing such corridors would also provide a

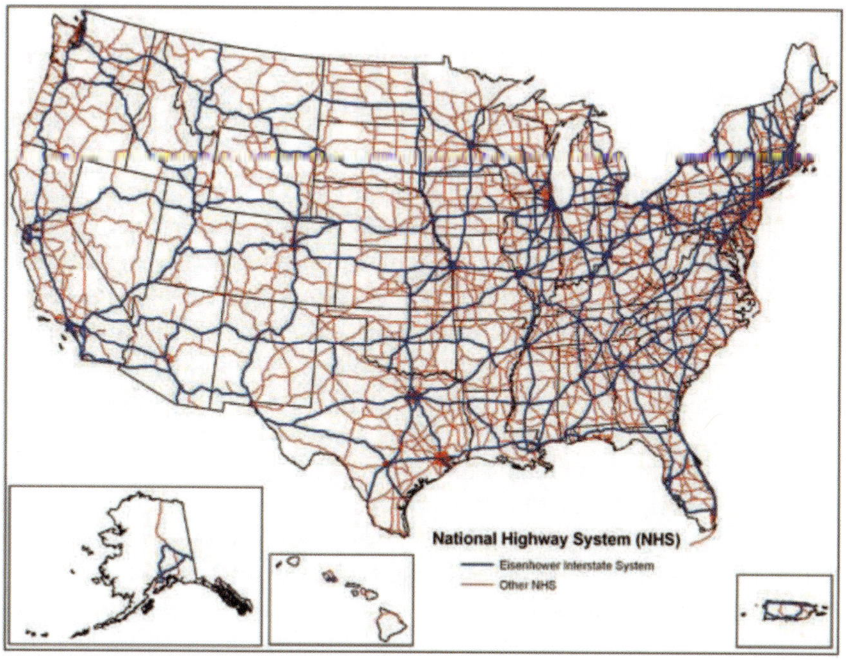

FIGURE 5-1 Map of the National Highway System. SOURCE: U.S. Department of Transportation.

logical tie to the proposed regional research centers as a means to foster cooperation in the road weather research program. Partnerships involving all key parties will develop naturally around these demonstration roadways.

The committee suggests that one north-south and one east-west interstate highway (Figure 5-1) be designated as national road weather demonstration corridors. These interstates should traverse areas with diverse weather conditions, topography, and geography. This will allow highway problems such as flooding, fog, evacuation, landslides, and winter weather to be addressed to a large extent. Criteria for selecting the corridors should include the extent to which instrumentation and equipment installed for other purposes along the selected interstate could be incorporated as part of the road weather demonstrations, as well as the willingness of state departments of transportation located along the route to be active participants. Further, the corridors should pass through several states that are early adopters of intelligent transportation system technology and are agreeable to serving as leaders in the implementation of the national road weather research program. For example, Interstate 35, a major North American Free Trade Agreement highway connecting Canada, the United States, and Mexico, may be a good candidate for a designated corridor; it traverses several climate zones and offers opportunities for early success because several states along this route already have invested in intelligent transportation system technology.

Various transportation research and development efforts, including the Strategic Highway Research Program, have employed demonstration projects, and some intelligent transportation system research is being implemented and tested with demonstration projects. The demonstration corridor concept is consistent with transportation research strategies previously used with success. These experiences have shown that successful implementation of results by a single state department of transportation can be extremely beneficial in gaining the support of other states. Therefore, the committee anticipates that the departments of transportation in each state through which the demonstration corridors pass could serve a role similar to that of the lead states in the Strategic Highway Research Program.

The committee foresees the road weather research program as providing seed money to develop these national corridors and then evaluate the results of the demonstrations. Actual implementation of new technologies would likely cost significantly more. The committee expects that costs associated with buying, installing, and operating technologies to be showcased would likely be covered by the states involved in the program, using other federal and state funds. Once the corridors are established, some of this funding would go to research to evaluate the feasibility and cost-effectiveness of implementing the demonstration technology more widely.

Establish Dedicated Road Weather Research Funding in the FHWA

Transportation research and management are currently highly decentralized, largely implemented by states, and meteorological research and operations are spread across several federal agencies, universities, and research centers. The private sector has provided many of the targeted road weather services to date. Yet, for the proposed road weather research program to succeed, centralized leadership at the federal level is essential in setting research priorities, administering grants, providing a central repository for research findings, ensuring accountability, and fostering the transfer of new knowledge and technology to operational practice. The Federal Aviation Administration's Aviation Weather Research Program provides a good model for the federal leadership needed for a road weather research program. The committee recommends that the FHWA have the leading role in the national road weather research program and that the National Oceanic and Atmospheric Administration (NOAA) be a lead partner in this effort. The federal government should establish an interagency coordinating council to guide the national road weather research program. In addition to the FHWA, participation on this council should include at a minimum NOAA and the National Science Foundation, especially through their joint U.S. Weather Research Program. It is no less critical that the leadership of the proposed road weather research program foster effective partnerships with the private sector in a manner that provides clear processes for interaction.

The committee recommends that long-term dedicated funding for road weather research sufficient to achieve the program's goals be established as *new* funding in the FHWA. The committee estimates that the research program will require on the order of $25 million per year based on an assessment of unmet needs, the capacity of the research community to conduct the program effectively, and costs for comparable transportation research initiatives in other areas, such as the Aviation Weather Research Program. The committee believes that many of the research and development challenges that have been identified will require more than a decade to solve, thus a long-term research and development program with appropriate funding commitments is required.

Although the FHWA is recommended as the primary funding recipient, the committee recommends that the dedicated funds for this program be highly leveraged with other initiatives that have an affinity with road weather research and development. The Aviation Weather Research Program is a good example; most of the focused research tracks in that program have produced results that can be integrated into the road weather program. The new funds will provide the necessary resources to focus both new research and the leveraged research from other programs to address unmet needs in the surface transportation community. Ongoing research sponsored by NOAA and the National Science Foundation and research in the U.S. Weather Research Program (about $60 million in FY 2002) can be leveraged similarly. This new FHWA funding would be a small increment to the current federal support for either transportation research (about $400 million, see Table 5-1) or meteorological research (about $600 million, see Table 5-2).

Fostering a Vigorous Private Sector

In addition to the need for an immediate infusion of federal funding, public-private-academic partnerships are needed to take up the research challenge, but this will be complicated by the several components of the public sector (federal, state, regional, and local) that will need to cooperate to achieve optimum results. This issue of public-private-academic partnerships in meteorology has received increasing attention in recent years. The National Research Council recently completed a study, *Fair Weather: Effective Partnerships in Weather and Climate Services* (NRC, 2003b), which found that boundaries in responsibilities between the public sector, private sector, and academia are not rigid. In general, each sector has played the following roles:

- The public sector is responsible for protecting life and property and enhancing the national economy. To carry out its mission it maintains an infrastructure of observation, communications, data processing, and prediction systems and conducts research on which the public, private, and academic sectors rely.
- Academia is responsible for advancing science and educating future meteorologists.
- The private sector is responsible for creating products and services tailored to the needs of its companies or clients and for working with the public sector to communicate forecasts and warnings that may affect public safety.

This three-sector system has led to extensive and flourishing weather services that greatly benefit the U.S. public and the economy. The report concluded that it was counterproductive and diversionary to establish detailed and rigid boundaries for each sector that specify who can do what and with what tools. Instead efforts should focus on improving the processes by which the sectors interact. These lessons clearly apply to the roles of government, academia, and the private sector in pursuing road weather research and development.

The committee recommends that steps be taken to foster effective public-private-academic partnerships in the area of road weather research and technology implementation. Essential partners in this effort include FHWA, NOAA, National Science Foundation, American Association of State Highway and Transportation Officials, academia, state and local governments, the private sector, and nongovernmental organizations and associations such as the Intelligent Transportation Society of America and the American Meteorological Society.

The committee believes that the preceding recommendations concerning regional research centers, a focus on national demonstration corridors, and widely accessible databases will facilitate such partnerships. Indeed, the contributions of all sectors are essential to a successful implementation of the program recommended here. The committee endorses the recommendations of *Fair Weather: Effective Partnerships in Weather and Climate Services* (NRC, 2003b) as they would apply to the road weather research program. Well-coordinated research and operational implementation objectives in the public sector will focus the efforts of the academic sector and provide the private sector with increased opportunities to improve their products and services and thus grow their industry.

Research Program to Be Guided by Measures of Cost-Effectiveness

The research program described in this document should result in a great number of research findings that can be developed further into new road weather applications (products, services, tools, and technologies). Indeed, multiple potential solutions may be developed to address a particular aspect of a road weather problem. Thus, it is important that decisions about which applications to develop and then deploy widely be guided by careful cost-benefit analyses. This cost-benefit information must be presented in ways that can be easily used by decision makers in allocating scarce resources. In assessing cost-effectiveness, a fairly broad definition of costs and benefits should be considered; for example, in addition to considering those costs and benefits that can be directly quantified, other indirect factors, such as environmental impacts that might result from reducing the amount of chemicals that drain off the roadway, must be considered.

As an example of how cost-effectiveness analyses can be useful, this report recommends that additional pavement sensors may be needed to provide the level of information necessary to make informed decisions about winter roadway maintenance. However, before installing large numbers of new sensors, research should be done to assess the marginal benefits and costs associated with adding more sensors as well as how those costs and benefits differ based on such factors as geographic location. A result of this research might be a methodology for determining the optimal number and location of sensors.

TECHNOLOGY TRANSFER

It is one of the ironies of federally sponsored research that many promising results are not effectively implemented in practical applications. Many results that do make it into practice take seemingly inordinate amounts of time to do so (NRC, 2000). For example, about 100 research products resulted from the Strategic Highway Research Program, a $150 million, five-year research program. A few of the lesser products, such as new signing concepts for work zones, were put into practice relatively quickly by several highway agencies and private sector contractors. The major subject of this research, improved asphalt pavements, has taken many years to implement across the nation. This slow application of research results is due in part to the lack of provisions to foster implementation at the state and local levels in the original program design. It was generally assumed that highway agencies would see the immediate benefits of the research results and simply implement them without any external stimulus. While some of this happened, years of effort by the FHWA, state departments of transportation, the American Association of State Highway and Transportation Officials (AASHTO), and the private sector have been required to produce meaningful impacts across the nation. In the case of the major asphalt research results, implementation was finally brought about by identifying lead states and assembling producer-user groups in states and regions that focused on implementation. AASHTO proceeded to develop and adopt new specifications, and the FHWA provided strong support and leadership to the project.

This report is focused on identification of research required for the surface transportation community. As illustrated by the Strategic Highway Research Program asphalt example, an essential component of the overall research effort is the transfer of results to operational users in an effective and timely manner. No matter how potentially valuable the research findings may be, if the resulting technology is not effectively transferred from the research community, this value is lost to stakeholders working in the roadway environment. In the road research community one reason for the research-to-operations barrier for the surface transportation community is that research, procurement, and operations often are disconnected in a transportation agency. Thus, the needs of the end user are seldom considered until moving into the implementation phase. Once a project has reached this point, change becomes so difficult and expensive that it is discouraged. Lack of any sense of ownership by the end users, who are often the ones responsible for making the technology work in the field, leads to spotty, slow implementation and consequently a low benefit-cost ratio. The old "failure to plan is planning to fail" adage could be rewritten to read "failure to fully understand and respond to user needs, abilities, and prevailing culture will result in failure to implement."

Early and continuing engagement of user or stakeholder communities throughout the research and development process is a critical aspect of technology transfer. "End-to-end" research and development sometimes is used to describe the identification of research needs, conducting the research, presenting the research for peer review, developing identifying applications from the research results, and finally deploying the applications and validating them in an operational environment. Unfortunately the last step is often the most problematic primarily because so few organizations have learned to do it well. The committee has identified several elements of effective transfer that have been validated over many decades.

- *Early and Continuing Involvement of Operational Stakeholders:* Operational users of the technology should be made a part of the technology transfer team from the beginning. Their continuing involvement ensures that the transfer path is efficient, expeditious, and relevant to the needs of the operational community.
- *Highly Resolved User Requirements:* Requirements for use of the new technology in an operational environment must be defined with as much resolution as possible to ensure effective implementation. This process includes time horizons for key operational decisions, specific weather-related information required for decision making, cost-benefit analysis of the weather-related decisions made by the user, level of technical competence of the users, training requirements, and documentation requirements. The more that is known about the user's operational processes, the more effective the transfer will be.
- *Iterative Development Process:* An iterative process that can be described as an ongoing dialogue between the development team and the user community characterizes successful transfers of technology. Early prototypes are based on a careful analysis of the user's requirements. These prototypes are evaluated and modified in several iterations to continually improve the performance of the technology. Improvements cease when cost of further iterations is judged to exceed value derived from the next change.
- *Validation in an Operational Setting:* It is essential to validate the new applications or technology in operational settings that represent all the modes of operation that the user may experience. This is often a step that is left out of the transfer process and results in a failed attempt to transfer the

technology since problems that surface only in an operational environment are not detected and addressed during the validation phase.

- *Comprehensive Training and Documentation:* Successful technology implementation over the long-term requires training for all users and educational programs for those in the operational community who will maintain and upgrade the technology. Complete documentation for each group of users is also essential.

Recommendation: Enable efficient technology transfer.

The committee recommends that the national road weather research program involve operational users as part of the research and development effort from the beginning. The user community for road weather information is broad and includes the general public, commercial shipping and bus transit companies, emergency managers, those who build, maintain, and operate the nation's roadways, and vehicle manufacturers. Early user involvement helps to ensure that the technology transfer path will be efficient and expeditious, and will produce results relevant to the needs of the user community.

EDUCATION AND TRAINING OF ROAD WEATHER INFORMATION USERS

User education programs should be developed as the research proceeds (1) to build community between researchers and users through feedback; (2) to develop constituencies supporting deployment of new technologies; and (3) to ensure that road users gain sufficient benefit from deployed road weather information systems to justify the expense. The proposed road weather research program should include support for this type of outreach activity. This section provides a brief discussion about the need to educate those who are the targeted users of road weather information. Many very diverse users of road weather information, including the traveling public, commercial vehicle operators, road maintainers, and emergency vehicle operators, must be educated on the value of using road weather information in their decision making and then trained on how to use the new technologies and informational tools as they are deployed.

The focus of the following sections is on training of road weather information *users;* however, training for the *providers* of this information is no less important. New professionals in the road weather industry must be trained on the job because they do not gain the necessary skills in either undergraduate or graduate academic programs. Including road weather in the curriculum used to train meteorologists and transportation researchers could go far toward building a more robust road weather research community (Osborne, 2001). Increasing support for road weather research at universities, a likely outcome of the proposed program, will create opportunities for graduate and undergraduate students to participate in such research projects as part of their education.

Education and Training of the Driving Public

To justify the expense of developing and deploying a road weather information system, the public must prepare for and then properly react to inclement weather encountered while on the roadways, purchase vehicles incorporating new technologies that enhance response to weather hazards, and access and use specialized weather data in decisions involved in trip planning, route selection, and safe vehicle operation. For these things to occur the driving public must receive proper education and training on road weather information and how to apply it. Such training begins with new drivers, who typically receive training through a combination of interactive instructional materials and hands-on modules. Instructional materials may include some weather modules, but training on driving in inclement weather is not mandatory nationwide. A national effort to develop training modules for weather-degraded road conditions could yield improvements in highway safety. Development of such modules should take advantage of experts in weather, in the physics of highway transit and automotive response, and in human factors to maximize their efficacy.

The driving public also needs to be continually updated about and trained to use new tools and technologies designed to improve vehicle performance under inclement weather; for instance, the anti-lock brake system is a safety feature now available in many new car models, but veteran drivers may not be aware of the difference between this feature and the classic braking system, nor may they be aware of how to use the system properly. Thus, informative manuals and hands-on training need to be a fundamental component of all new tools and technologies as they become available. Veteran drivers may need supplemental training if they move or travel to regions where they encounter weather and road conditions not previously experienced. For example, someone who travels from Miami to Minneapolis in winter may not know how to handle snowy or icy roads, and thereby may pose a threat to themselves and others. Rare weather conditions, such as snow in the Southwest, or rapidly changing situations, such as a cloudburst, ground fog, blowing dust, or even the first snow of the season, may also pose dangers. Opportunities for continuing training could help the driving public learn how to properly respond to these situations.

A wealth of weather information is available to roadway users today and even more will be available in the future, but drivers that know what information is available and how to access and use it are in the minority. Many know that weather nowcasts and forecasts are available on commercial radio and television but are unaware of other sources, such as from Web sites or 511 (the telephone number for receiving traveler information, ultimately to be available nationwide), that provide more detailed data tailored to their needs. Effective outreach to the driving public is needed to inform them about new road weather information and tools. As traveler weather information services evolve the public must become aware of the opportunity to keep informed about the driving environment and understand their obligation to respond accordingly.

Drivers of commercial vehicles used for long-haul shipping or just-in-time delivery will need specialized training on how to use enhanced road weather information to increase safety and efficiencies (and hence a competitive edge) for their companies. This training must address both how to best operate vehicles in inclement weather and how to interpret weather information and use it to make optimal decisions. Parallel complementary training for dispatchers, who make choices about routes and timing for deliveries, will also be necessary to gain full benefit.

An output of a long-term road weather research program will be the development of "intelligent agents" that will be embedded in "smart" vehicles and personal information management systems (the successors to today's personal data assistants). Such devices will automatically access the necessary weather data and inform their users in tailored ways that reflect the user preferences and needs, automatically suggest routes and itineraries, and free the driver to focus on vehicle operation. The development of such sophisticated personal tools is unlikely to reduce the need for education and training—it will just change their nature.

Education and Training of Roadway Maintainers and Operators

A recent study to identify weather information needs for surface transportation showed that within any given surface transportation sector, all users did not clearly understand how weather information could make a positive, significant difference in their operations (OFCM, 2002a). The study also revealed a polarization in user sophistication in using weather data and in comprehension of how better information could improve operations.

These findings speak to the need for training of roadway maintainers and operators on how to use enhanced weather information. Such training will be critical for successful implementation of improved road weather technologies and services. Because many meteorological phenomena are highly seasonal, annual refresher training will be important. Technology now allows many decision support systems to incorporate just-in-time refresher training, essentially coaching a user through an unfamiliar situation. To be most effective, educational programs should allow feedback from the users, enabling evaluation of new road weather products and information and allowing assessment of the effectiveness of the instruction.

State, county, and municipal employees responsible for road maintenance and construction, emergency management, law enforcement, and traffic management will need training as new road weather technologies and tools are deployed. In particular, equipment operators often have little or no exposure to emerging technology, and therefore may view it as a threat to their jobs rather than as a tool to help them perform their work more effectively. Introducing new information and technology to them must be done incrementally, providing extensive but practical training with user-friendly displays in a way that allows all workers to understand how their efforts contribute to solving the overarching problem.

A number of mechanisms are available for training transportation professionals on new technologies and tools, including pilot projects, operational demonstrations, and tabletop exercises,[2] all of which allow the users to test out the new technologies in a real-world setting. Computer training modules are also emerging as an effective way to train roadway maintainers and operators. This approach allows for appropriate and understandable information to be disseminated in a timely manner and can provide a user-friendly, interactive training interface. Such modules can be made available on a CD, or the Web, which allows more frequent updates. In fact, a Web site could be staffed by people who could help users navigate the site and answer questions pertinent to the users' immediate needs (see http://www.comet.ucar.edu/modules/).

Training and education of those responsible for maintaining and operating the nation's roads will likely be performed by various public and private

---

[2]Tabletop exercises are simulated events that allow emergency managers to practice how they would coordinate their efforts to respond effectively. These exercises are used for rare events, such as hurricanes, chemical spills, or terrorist attacks, that require rapid response by multiple jurisdictions and emergency management entities.

organizations. Federal oversight would be useful in terms of interagency coordination, facilitating transfer of knowledge from one state to another, and identifying user training needs. Because the private sector likely will play an important role in developing practical applications of road weather information, it also should be involved in education and training of surface transportation professionals.

Recommendation: Improve education and training of road weather information users.

The road weather research program should incorporate training on the use of road weather information and technology into driver education nationwide. This training needs to start with new-driver education and be reinforced periodically as new technology is introduced or other circumstances arise, such as when a driver moves to a new region with a different climate. In commercial settings continuing education on road weather information is recommended for drivers and dispatchers. Special efforts are needed in the education and training for such groups such as road maintainers, emergency managers, and traffic managers. When new products or decision support tools are deployed, educational and outreach programs, pilot projects, operational demonstrations, and "table-top exercises" should be conducted on a regional basis.

RESEARCH SYNERGIES AND EFFICIENCIES WITH OTHER MODES OF TRANSPORTATION

The federal government currently supports numerous research efforts on the impact of weather on other modes of transportation, including aviation, rail, and transit. Some of these projects investigate micrometeorological problems near the surface. Others have developed tools and techniques that may be applicable to road weather research. Given the limited fiscal resources available to support road weather research, it is important that the road weather research community be knowledgeable of the research being conducted on other modes of transport, and adopts, adapts, and applies the results of that research where applicable. Where there is common interest in a meteorological phenomenon, such as fog, the road weather research community should collaborate with researchers in the aviation weather community. It is also likely that important topics in road weather research

FIGURE 5-2 Airport runway snowplow. SOURCE: Schmidt Engineering and Equipment, Inc.

will have applicability to other modes; for example, the removal of ice and snow from hard-surfaced lanes is applicable to frozen railway switches and to clearing of snow and ice from airport runways and along city transit routes (Figure 5-2). Likewise, research on how to detect and predict surface visibility is applicable to all modes of transportation.

Each transportation mode also conducts research and development focused on tactical and strategic decision support systems. Techniques for building these tools, stakeholder interaction, and operational deployment are needed across all modes of transportation. Some situations, such as evacuations caused by hurricane landfall or spread of hazardous materials, require coordination of roadway, air, rail, and transit modes, making it appropriate to consider all modes when conducting this sort of research.

Lessons can be learned from aviation weather research, which has been conducted for decades, and has much applicability to all surface transportation modes. The Federal Aviation Administration's Aviation Weather Research program has two decades of research results in thunderstorm forecasting, icing, visibility, decision support systems, and crosswinds. Other transportation modes can clearly make use of the accomplishments of this research. In addition, lessons can be learned from the aviation weather experience about how to build an interdisciplinary research program with close ties to operational applications. Given the large investment in aviation weather that the nation has made in the last decade or so, learning from and collaborating with the aviation research community should be an early step in the development of a national road weather research program.

Recommendation: Seek out synergies and efficiencies between road weather research and parallel efforts regarding other modes of transportation.

The committee recommends that the road weather research program routinely monitor research parallel to surface operations in aviation, rail, and transit to exploit applicable overlap. Many aspects of the proposed road weather research program could benefit research in aviation, rail, and other modes of transportation. Indeed, there are many synergies and efficiencies to be gained by coordinating research on meteorological phenomena that affect all modes and on the development of decision support resources. In particular the well-established aviation weather research programs offers many lessons learned that should be examined by the emerging road weather research program.

6

Closing Thoughts

Amy is the senior traffic manager for the Dallas–Fort Worth Metroplex area, the "Golden Triangle" that includes Denton to the north, Dallas to the southeast, and Fort Worth to the southwest. It is November 25, 2030, the Wednesday before Thanksgiving, and the American Automobile Association is predicting that 37 million people will be traveling 50 miles or more by car; of those, 750,000 are expected in the Dallas–Fort Worth area, and many have already begun to hit the roads. In her operations center Amy watches a late-season line of strong thunderstorms move through the area accompanied by lightning, heavy rain on the order of an inch per hour, and winds gusting to 40 mph. Her radar display indicates the heaviest precipitation has moved into the metro area. Amy consults her traffic simulation model that monitors traffic and weather in real-time and predicts congestion out to three hours; she sees that there should be no major delays for the millions of people trying to get to their friend's or relative's home to eat turkey and give thanks. Her suite of traffic cameras confirm the model; upon onset of the heavy rain, all vehicles have automatically slowed and reconfigured for optimal traction on wet roads, and the automatic spacing between the cars has increased. Although the line of storms is narrow, Amy thinks back to earlier in this century when a line of storms similar to this would have jammed the already busy freeways and caused accidents, further compounding the holiday congestion and creating ripple effects that would have lasted throughout the day. Thankful for the research and technology that now facilitate smooth traffic flow in inclement weather, Amy sits back and relaxes, unworried by the steady rhythm of rain on the roof of the building.

Now is the time for a focused research effort to be undertaken that will allow the road weather system of the future to be realized. An opportunity exists to leverage the knowledge from analogous programs, the efforts from existing projects (e.g., intelligent transportation systems, the Maintenance Decision Support System), and the expertise of the meteorological and transportation communities in order to improve the safety and efficiency of a transportation system that is used by millions of people every single day and is affected by weather every single day. Decision makers can help achieve the vision of the roadway system of the future portrayed above and in the earlier vignettes (see Chapter 2) by capitalizing on the framework provided in this report and bringing the road weather research program to fruition. Such a program would be the first of its kind; coming just over 100 years since the automobile was invented, it is a program that is long overdue.

The advances in knowledge and operational capabilities that could result from a comprehensive road weather research program such as that proposed in this report have the potential to improve public safety, economic efficiency, security, and environmental quality for the nation. Improved road weather information for the driving public should reduce the number of weather-related accidents, and the associated fatalities, injuries, and property damage. At the same time, individual drivers and commercial trucking operations would be able to find better routes that minimize exposure to weather threats, leading to less time spent on the road and more efficient shipping practices. Benefits also would be gained by developing weather information products and decision support tools that would enable a more cost-effective and optimized operation of the roadway system, from advanced winter maintenance practices to weather-responsive traffic management. Other advantages, such as the ability to better respond to chemical spills on the roadway or reduced environmental impacts from chemicals used to treat snow and ice on the road, are also likely (Ham and Lockwood, 2002). Although these benefits have not been quantified, they could be substantial if the proposed road weather research program were implemented to exploit the potential of current and new tools and technologies.

Beyond these direct benefits of improved road weather information and services, an enhanced understanding of how to address road weather problems will position the nation to better respond to many other large-scale socioeconomic and physical changes relevant to the roadway environment. Indeed, global trends in demographics, the environment, and technology will affect road transportation dramatically during the next decade. Thus, any improvements and efficiencies to be gained in the surface transportation system, for example, from more strategic responses to inclement weather, are desirable.

Changes in the demographics of Western industrialized nations include increased life expectancy, leading to both an increasing and an aging population. These changes are compounded by the fact that relatively few new roadways are being built to accommodate the increasing volume of vehicles on the roadways, and the major road construction changes that are being undertaken often take on the order of 5 to 10 years to complete. An additional demographic change is the increase in the number of nonnative English speakers and non-English-speaking individuals. The proposed road weather research program will help alleviate some of the difficulties of these demographic changes with better traffic management, improved weather forecasting, human factors research, and vehicle telematics.

The population distribution is another demographic trend that will require attention over the next decades. Per capita vehicle ownership continues to increase while populations continue to disperse away from major urban centers. The result of these trends is ever greater volumes of vehicles to carry commuters to and from work. For example, from 1980 to 1990, 22 million drive-alone workers were added to the roads while car-pooling rates dropped sharply (NRC, 1998). The high speeds and close spacing of vehicles on the approaches to major urban centers provide the ingredients for accidents involving large numbers of vehicles as soon as conditions deteriorate. Indeed, the last decade has seen an increasing number of very large multivehicle accidents, sometimes involving on the order of a hundred vehicles, brought on by a sudden drop in visibility due to fog or in traction due to ice and snow. Integrated intelligent transportation systems solutions have been devised that can reduce these risks substantially, for instance, with on-vehicle safety applications that enable closer spacing and automated braking. However, mass production of affordable vehicles with this technology is some time away, and it will be longer still before a sufficient proportion of vehicles are equipped.

Environmental considerations have preoccupied the road transportation community for decades already, resulting in large reductions in vehicle emissions that cause air pollution. In response to global climate change, the transportation community is making additional technological changes in vehicles to reduce emissions of greenhouse gases, largely through the development of hybrid vehicles and hydrogen fuel cells. However, hydrogen fuel cells emit only water vapor, which could substantially impact the micrometeorology of the roadway environment with fog development or formation of black ice on the roadway. The proposed road weather research program will facilitate the observation and modeling of the microscale roadway environment, which will improve understanding of fog and ice formation in the complex roadway environment, and it will help roadway operators and maintainers manage these conditions with improved forecasting and vehicle telematics.

Technological solutions from the intelligent transportation systems community will provide many of the required future solutions. Open systems and common standards for vehicle systems and the road infrastructure across the entire continentwide road system will help North America remain competitive in the global marketplace. But, technology alone will not be enough to attain the full benefits of improved road weather information and services. Transportation professionals will need to continue to move more aggressively from maintaining the system to managing it proactively, and using current and forecast weather data to anticipate the effects of weather on traffic flow is absolutely necessary to do that. Very careful planning and management will be required to ensure effective application of the intelligent transportation systems, sophisticated meteorological solutions, and other technologies. Part of that effort will have to go toward planning on a longer timescale to assess the potential impacts of these and other global trends. Longer-timescale problems require longer lead times to solve, making their early detection all the more crucial. Careful planning combined with focused road weather research will help the nation overcome the transportation challenges of the coming decades.

References

Andrey, J., J. Li, and B. Mills. 2003a. A winter index for benchmarking winter road maintenance operations on Ontario highways. 80th Annual Meeting Preprint CD-ROM, proceedings of the Transportation Research Board annual meeting, Washington, D.C., January 7-11, 2001.

Andrey, J., B. Mills, M. Leahy, and J. Suggett. 2003b. Weather as a chronic hazard for road transportation in Canadian cities. Nat. Hazards 28:319-343.

Badger, J. 1996. Human factors. Harris Technical Services. Available at http://www.harristechnical.com/human.htm. Accessed October 24, 2003.

Boselly, S. E. 2001. Benefit/cost study of RWIS and anti-icing technologies. National Cooperative Highway Research Program Report No. 20-7(117).

Cambridge and Mitretek (Cambridge Systematics, Inc., and Mitretek Systems, Inc.). 2003. Weather in the infostructure. Available at http://www.its.dot.gov/511/PDF/wethinfo.pdf. Accessed December 6, 2003.

Chrysler, S. T., S. M. Danielson, and V. M. Kirby. 1997. Age differences in visual abilities in nighttime driving field conditions. In Designing for an Aging Population: Ten Years of Human Factors/Ergonomics Research, W.A. Rogers, ed. Santa Monica, CA: Human Factors and Ergonomics Society.

Collins, D. J., V. L. Neale, and T. A. Dingus. 1999. Driver performance when using an in-vehicle signing information system considering adverse weather, visibility condition, and age. Proceedings of the Intelligent Transportation Society of America 9th Annual Meeting and Exposition (CD-ROM). Washington, DC: Intelligent Transportation Society of America.

Crevier, L.-P., and Y. Delage. 2001. METRo: A new model for road-condition forecasting in Canada. J. Atmos. Sci. 40:2026-2037.

Eriksson, M. 2001. Regional influence on the occurrence of road slipperiness during winter precipitation events. Meteorol. Appl. 8:449-460.

FHWA (Federal Highway Administration). 2002. An Introduction to Standards for Road Weather Information Systems (RWIS): Siting Standards, Calibration Standards, Communication Standards. Publication No. FHWA-OP-02-079.

FHWA (Federal Highway Administration). 2003. Road weather management. Available at http://www.ops. fhwa.dot.gov/Weather. Accessed October 24, 2003.

Franklin, J. L., C. J. McAide, and M. B. Lawrence. 2003. Trends in track forecasting for tropical cyclones threatening the United States, 1970–2001. Bull. Am. Meteorol. Soc. 84:1197-1203.

Geiger, R., R. H. Aron, and P. E. Todhunter. 1995. The Climate Near the Ground. Fifth Edition. Wiesbaden, Germany: Vieweg.

Gish, K., L. Staplin, and M. Perel. 1999. Human factors issues related to the use of vision enhancement systems. Transportation Research Record, 1694.

Glahn, H. R., and D. P. Ruth. 2003. The new digital forecast database of the National Weather Service. Bull. Am. Meteorol. Soc. 84:195-202.

Goodwin, L. 2003a. Best practices for road weather management, Version 2.0. Prepared for the Federal Highway Administration (FHWA) Road Weather Management Program. Available at http://www.ops.fhwa.dot.gov/weather/best_practices/CaseStudiesFINALv2-RPT.pdf. Accessed January 27, 2004.

Goodwin, L. 2003b. Weather-related crashes on U.S. highways in 2001. Available at http://www.ops.fhwa.dot.gov/weather/docs/2001CrashAnalysisPaperV2.doc. Accessed January 27, 2004.

Goodwin, L., and P. Pisano. 2002. Current practices in transportation management during inclement weather. Institute of Transportation Engineers Annual Meeting, Philadelphia, Pennsylvania, August 4-7, 2002.

Gray, W. M., C. W. Landsea, P. W. Mielke Jr., and K. J. Berry. 1993. Predicting Atlantic basin seasonal tropical cyclone activity by 1 August. Weather Forecast., 8:73-86.

Hallowell, R. 2003. Personal communication, June 24, 2003.

Ham, D. B., and S. Lockwood. 2002. National needs assessment for ensuring transportation and infrastructure security. National Cooperative Highway Research Program Project 20-59, Task 5. Available at http://www.transportation.org/community/security/doc/NatlNeedsAssess.pdf.

Horel, J., T. Potter, L. Dunn, W. J. Steenburgh, M. Eubnk, M. Splitt, and D. J. Onton. 2002a. Weather support for the 2002 Winter Olympic and Paralympic games. Bull. Am. Meteorol. Soc. 83:227-240.

Horel, J., M. Splitt, L. Dunn, J. Pechmann, B. White, C. Cilberti, S. Lazarus, J. Slemmer, D. Zaff, and J. Burrs. 2002b. MesoWest: Cooperative mesonets in the western United States. Bull. Am. Meteorol. Soc. 83:211-225.

ICBC (Insurance Corporation of British Columbia). 2000. Traffic Collision Statistics. Available at http://www.icbc.com/Library/research_papers/Traffic/. Accessed January 28, 2004.

Jacobs, W., and W. E. Raatz. 1996. Forecasting road-surface temperatures for different site characteristics. Meteorol. Appl. 3:243-256.

Kalnay, E., S. Lord, and R. McPherson. 1998. Maturity of operational numerical weather prediction: Medium range. Bull. Am. Meteorol. Soc. 79:2753-2769.

Lombardo, L. 2000. Overview of US Crashes and Environment. Presentation at the WIST II Forum, December 4-6, 2000.

Mahoney, W.P., and W. Meyers. 2003. The Winter Road Maintenance Decision Support System (MDSS) Project Update and Future Plans. Presentation at the American Meteorological Society Meeting, Long Beach, California, February 9-13, 2003.

Mass, C.F., M. Albright, D. Owens, R. Steed, M. MacIver, E. Grimit, T. Eckel, B. Lamb, J. Vaughan, K. Westrick, P. Storch, B. Colman, C. Hill, N. Maykut, M. Gilroy, S. A. Ferguson, J. Yetter, J. M. Sierchio, C. Bowman, R. Stender, R. Wilson, W. Brown. 2003. Regional environmental prediction over the Pacific Northwest. Bull. Am. Meteorol. Soc. 84:1353-1366.

Miller, P. A. and M. F. Barth. 2003. Ingest, integration, quality control, and distribution of observations from state transportation departments using MADIS. International Conference on Interactive Information and Processing Systems for Meteorology, Oceanography, and Hydrology (IIPS), Feb. 9-13, 2003.

Miller, R. 2003. In-vehicle telematics. Presentation to the Committee on Weather Research for Surface Transportation: The Roadway Environment. Seattle, WA. April 24, 2003.

Mimbela, L. Y. and A. L. Klein. 2000. Summary of vehicle detection and surveillance technologies used in Intelligent Transportation Systems (ITS), Vehicle Detector Clearinghouse (VDC). Available at http://www.fhwa.dot.gov/ohim/tvtw/vdstits.pdf. Accessed January 27, 2004.

Moninger, W., R. Mamrosh, and P. Pauley. 2003. Automated meteorological reports from commercial aircraft. Bull. Am. Meteorol. Soc. 84:2, 203-216.

National ITS Architecture Development Team. 2003. National ITS Architecture Executive Summary. Available at http://www.itsdocs.fhwa.dot.gov/jpodocs/repts_pr/45_01!.pdf. Accessed October 28, 2003.

NRC (National Research Council). 1994. Toward a New National Weather Service—Weather for Those Who Fly. Washington, D.C.: National Academy Press.

NRC. 1998. Atmospheric Change and the North American Transport Sector: Summary of a Trilateral Workshop. Washington, D.C.: National Academy Press.

NRC. 2000. From Research to Operations in Weather Satellites and Numerical Weather Prediction: Crossing the Valley of Death. Washington, DC: National Academy Press.

NRC. 2002. Weather Radar Technology Beyond NEXRAD. Washington, DC: The National Academies Press.

NRC. 2003a. Communicating Uncertainties in Weather and Climate Information. Washington, DC: The National Academies Press.

NRC. 2003b. Fair Weather: Effective Partnerships in Weather and Climate Services. Washington, DC: The National Academies Press.

NRC. 2003c. Tracking and Predicting the Atmospheric Dispersion of Hazardous Material Releases: Implications for Homeland Security. Washington, DC: The National Academies Press.

NSF (National Science Foundation). 2002. Summary of FY 2002 Budget Request to Congress. Available at http://www.nsf.gov/bfa/bud/fy2002/start.htm. Accessed January 22, 2004.

NSF. 2003. Summary of FY 2003 Budget Request to Congress. Available at http://www.nsf.gov/bfa/bud/fy2003/start.htm. Accessed January 22, 2004.

NSF. 2004. FY 2004 Budget Request to Congress. Available at http://www.nsf.gov/bfa/bud/fy2004/toc.htm. Accessed January 22, 2004.

Nelson, R. J. 2002. Weather-based traffic management applications in Nevada. Institute of Transportation Engineers 2002 Annual Meeting, Philadelphia, Pa.

Nilson, L., and H. Alm. 1996. Effect of a vision enhancement system on drivers' ability to drive safely in fog. In Vision in Vehicles V, 263-271, A. G. Gale, ed. Amsterdam: North-Holland/Elsevier.

Nixon, W. A. 2001. Research needs in weather information for surface transportation—the perspective of the user community. Eleventh Annual ITS America Meeting, Miami Beach, Fla., June 4–7, 2001.

OFCM (Office of the Federal Coordinator for Meteorological Services and Supporting Research). 2001. The Federal Plan for Meteorological Services and Supporting Research: Fiscal Year 2002. Silver Spring, MD: U.S. Department of Commerce/National Oceanic and Atmospheric Administration. Available at http://www.ofcm.noaa.gov/fp-fy02/fedplan.htm. Accessed January 22, 2004.

OFCM. 2002a. The Federal Plan for Meteorological Services and Supporting Research: Fiscal Year 2003. Silver Spring, MD: U.S. Department of Commerce/National Oceanic and Atmospheric Administration. Available at http://www.ofcm.noaa.gov/fp-fy03/fedplan.htm. Accessed January 22, 2004.

OFCM. 2002b. Weather Information for Surface Transportation National Needs Assessment Report. Silver Spring, MD: U.S. Department of Commerce/National Oceanic and Atmospheric Administration.

OFCM. 2003. The Federal Plan for Meteorological Services and Supporting Research: Fiscal Year 2004. Silver Spring, MD: U.S. Department of Commerce/National Oceanic and Atmospheric Administration. Available at http://www.ofcm.noaa.gov/fp-fy04/fedplan.htm. Accessed January 22, 2004.

Oke, T. R. 1987. Boundary Layer Climates. Second Edition. London: Methuen and Co.

Olson, P. L., T. Aoki, D. Battle, and M. Flannagan. 1990. Development of a headlight system performance evaluation tool: Final report. Report No. DOT HS 807 697. Washington, DC: National Highway Traffic Safety Administration.

Osborne, L. F. 2001. Future growth of surface transportation weather: An academic question. Eleventh Annual ITS America Meeting, Miami Beach, FL, June 4–7, 2001.

Partanen, P., V. Haavisto, and T. Haavasoja. 2003. Laboratory and field test results from a pavement sensor. Eighty-third Annual American Meteorological Society Meeting, Long Beach, Calif., Feb. 9-13, 2003.

Perrin, J., P. T. Martin, and B. G. Hansen. 2001. Modifying signal timing during inclement weather. 80th Annual Meeting Preprint CD-ROM, proceedings of the Transportation Research Board annual meeting, Washington, DC, January 7-11, 2001.

Pielke, Jr., R. A., and C. W. Landsea. 1998. Normalized hurricane damages in the United States: 1925-95. Weather Forecast. 13:621-631.

Pisano, P., L. Goodwin, and A. Stern. 2002. Surface transportation safety and operations: The impacts of weather within the context of climate change. The Potential Impacts of Climate Change on Transportation Workshop, October 1-2, 2002. Available at http://climate.volpe.dot.gov/workshop1002/pisano.pdf or http://www.ops.fhwa.dot.gov/weather/best_practices/RWMP_Climate_Change_v3.pdf. Accessed January 27, 2004.

Postgard, U. 2001. Adjustment time for road surface temperature during weather changes. Meteorol. Appl. 8:397-407.

Prusa, J. M., M. Segal, B. R. Temeyer, W. A. Gallus, Jr., and E. S. Takle. 2002. Conceptual and scaling evaluation of vehicle traffic thermal effects on snow/ice covered roads. J. Appl. Meteorol. 41:1225-1240.

Rayer, P. J. 1987. The Meteorological Office forecast road surface temperature model. Meteorol. Mag. 116:180-190.

Reichardt, M. 2003. The Sensor Web's point of beginning: distributed devices that can measure physical perameters—such as temperature, air quality, or vibration—and report those measurements along with their position could revolutionize the way we discover and assess real-time data. Geospatial Solutions 13(4):40-43.

Sass, B. H. 1992. A numerical model for prediction of road temperature and ice. J. Appl. Meteorol. 31:1499-1506.

Smith, H. S. 2003. Impact of weather on pavement. Presentation to Committee on Weather Research for Surface Transportation: The Roadway Environment. Washington, DC, August 18, 2003.

Smith, J. 2001. Private sector meteorology and ITS. Eleventh Annual ITS America Meeting, Miami Beach, FL, June 4–7, 2001.

Spaid, J. 2004. Impacts of weather on highway construction. 84th Annual American Meteorological Society Meeting, Seattle, Washington, January 11-15, 2004.

Stokes, G. M., and S. E. Schwartz. 1994. The Atmospheric Radiation Measurement (ARM) Program: Programmatic background and design of the cloud and radiation test bed. Bull. Am. Meteorol. Soc. 75:1201-1221.

Stowe, R. 2001. A benefit/cost analysis of intelligent transportation system applications for winter maintenance. 80th Annual Meeting Preprint CD-ROM, proceedings of the Transportation Research Board annual meeting, Washington, DC, January 7-11, 2001.

SSI (Surface Systems, Incorporated). 2003. Pavement Sensors. Available at http://www.ssiweather.com/pages/products/rwis/pavement_sensors.htm. Accessed December 31, 2003.

Takle, E. S. 1990. Bridge and roadway frost: Occurrence and prediction by use of an expert system. J. Appl. Meteorol. 29:727-734.

TRB (Transportation Research Board). 1984. Special Report 202: America's Highways: Accelerating the Search for Innovation. Washington, DC: National Academy Press.

TRB. 1994. Special Report 251: Toward a Sustainable Future: Addressing the Long-Term Effects of Motor Vehicle Transportation. Washington, DC: National Academy Press.

TRB. 2001a. Special Report 260: Strategic Highway Research: Saving Lives, Reducing Congestion, Improving Quality of Life. Washington, DC: National Academy Press.

TRB. 2001b. Special Report 261: The Federal Role in Highway Research and Technology. Washington, DC: National Academy Press.

U.S. Census Bureau. 2000. Projections of the Total Resident Population by 5-Year Age Groups, and Sex with Special Age Categories: Middle Series, 2016 to 2020. Available at http://www.census.gov/population/projections/nation/summary/np-t3-e.txt. Accessed February 20, 2004.

USDOT (U.S. Department of Transportation). 2000. Advanced Traveler Information Service (ATIS): What do ATIS customers want? Available at http://www.itsdocs.fhwa.dot.gov/JPODOCS/REPTS_TE/9H801!.PDF. Accessed December 15, 2003.

USWRP (U.S. Weather Research Program). 2003. The U.S. Weather Research Program: National need, vision, and interagency plan for FY 2000-2006 (9/9/03 V4.2). Available at http://box.mmm.ucar.edu/uswrp/program_overview/USWRP_Program_Plan_4.2.pdf. Accessed October 24, 2003.

Vatalaro, M. 2001. Beat the jam. Tech. Rev. 104:36.

Wilhelmi, O. V., and J. C. Brunskill. 2003. Geographic information systems in weather, climate, and impacts. Bull. Am. Meteorol. Soc. 84:10.

Wojcik, G. S., and D. R. Fitzjarrald. 2001. Energy balances of curing concrete bridge decks. J. Appl. Meteorol., 40:2003-2025.

Wolshon, B., E. Urbina, and M. Levitan. 2001. National review of hurricane evacuation plans and policies. Baton Rouge: Louisiana State University Hurricane Center.

Wolshon, B., and M. Levitan. 2002. Evacuation route traffic, flood, and wind hazard monitoring system. Proceedings of the American Society of Civil Engineers—Solutions for Coastal Disasters Conference '02, San Diego, Calif., February 2002.

Appendixes

A

Statement of Task

This study will examine the research opportunities and required services needed to support improved weather-related information for the nation's roadways. It will investigate the current state of knowledge regarding road weather conditions (observed and predicted) and recommend key areas of research to enhance operational production of weather-related information for roads. In addition, it will identify possible agency requirements and the observational, computing, and communication infrastructure needed to best provide this information to users. It will have a practical focus, stressing not just research opportunities but how to make this information useful for improved operations and implementation. In essence, the study will provide a framework and recommendations to engage the transportation and weather communities (and other stakeholders) in the development of a strategic plan to guide road weather research and, where appropriate, surface transportation more generally. Topics to be addressed will include:

1. Briefly describe the nature and scope of road weather issues and general economic impacts.
2. Characterize the current state of road-related weather research, including status of ongoing Federal Highway Administration activities and other federal and state activities, status of activities within the meteorological community, and challenges presented by the interdisciplinary nature of the problems.
3. Identify gaps in the scientific knowledge base needed to support improved weather-related information for roadways and recommend research areas to increase understanding of weather specific to roads.

4. Discuss how research can be designed so it clearly supports operations and leads most efficiently to practical applications and discuss the scientific and technical feasibility of developing useful weather information.

5. Identify needs within the meteorological infrastructure that would be necessary to establish effective road weather research and operational capabilities.

B

Biographical Sketches of Committee Members and Staff

John T. Snow is the dean of the College of Geosciences at the University of Oklahoma and the director of the Oklahoma Weather Center Programs. Dr. Snow received a B.S.E.E. from Rose Polytechnic Institute in 1968, an M.S.E.E. from Rose Polytechnic Institute in 1969, and a Ph.D. from Purdue University in 1977. His research interests are in dynamics of geophysical columnar vortexes ranging in scale from small dust devils to fire whirls, with a focus on tornadoes; application of laser-Doppler velocimetry to laboratory measurements in experimental fluid mechanics; earth science education at all levels; meteorological measurements and instrumentation, especially surface instruments for weather observations; post-event analyses of tornadoes and tornado-producing thunderstorms; and experimental fluid mechanics as applied to atmospheric problems. He has served on the following National Research Council bodies: the National Council of Teachers of Mathematics Standards Review Committee, the Panel for Assessment of Wind Engineering Issues in the United States, and the Committee on Natural Disasters. Dr. Snow has served in several other advisory capacities, including as chair of the NEXRAD Technical Advisory Committee; ex-officio board member for Weather Decision Technologies, Inc.; chairman of the Board of Trustees of the University Corporation for Atmospheric Research; member and past chairman of the Board for Oceans and Atmosphere, National Association of State Universities and Land Grant Colleges; and commissioner for education and human resources of the American Meteorological Society.

Elizabeth Carter is president of WeatherExtreme, LCC, a consulting company that specializes in atmospheric, environmental, and snow research. WeatherExtreme is a certified contractor with the California Department of Transportation (CalTrans) and is currently working with the National Oceanic and Atmospheric Administration on establishing a snow-level product for CalTrans. Previously Dr. Carter was associate professor of mathematics, physics and meteorology at Sierra Nevada College in Incline Village, Nevada. She also was an associate scientist at the Atmospheric Sciences Center at the University of Nevada's Desert Research Institute. Dr. Carter received her Ph.D. in physics from the University of Nevada.

Dennis L. Christiansen has been a member of the staff of the Texas Transportation Institute for 30 years. He is presently deputy director of the institute and has overall responsibility for research on transportation operations, planning, and economics. The institute is currently undertaking a project to implement intelligent transportation systems to detect weather conditions in rural Texas. Dr. Christiansen has extensive research experience in several areas, including traffic operations, transportation planning, and transit planning. He has specialized in multimodal research with an emphasis on identifying cost-effective approaches to urban mobility concerns. His projects have addressed areas such as roadway operations and design; transportation and energy relationships; the potential role for a system of strategic arterial streets; and urban goods movement. Dr. Christiansen is recognized as an international expert in the planning, design, operation, and evaluation of preferential facilities for high-occupancy vehicles. He is a recipient of the Transportation Research Board's Fred Burgraff Award. He is presently a member of the Board of Direction for the International Institute of Transportation Engineers. He has also served on the Transportation Research Board's Committee for the Study of the Regulation of Weights, Lengths, and Widths of Commercial Motor Vehicles and currently serves on that board's Research and Technology Coordinating Committee.

Bradley R. Colman is the Science and Operations Officer at the National Oceanic and Atmospheric Administration (NOAA)/National Weather Service (NWS) Seattle–Tacoma Weather Forecast Office in Seattle, Washington, and an associate professor with the University of Washington's Department of Atmospheric Sciences. Dr. Colman's expertise is in operational weather analysis and forecasting, coastal and mountain meteorology, and numerical weather prediction. He has taught extensively within the operational community, particularly through the Cooperative Program for Operational Meteorology, Education and Training (COMET) in Boulder, Colorado. Prior to his current appointment with NOAA/NWS, Dr. Colman was a research meteorologist with NOAA's Forecast System Laboratory. Dr. Colman has recently co-edited a two-volume *Handbook of Weather Cli-*

mate and Water with Wiley Press. He has served on the Science Steering Team of the U.S. Weather Research Program. Dr. Colman is actively involved with the American Meteorological Society (AMS) where he has previously served as Chief Editor of the Weather and Forecasting journal. He has also served on the AMS Council and Executive Committee, and is currently Commissioner of the Scientific and Technological Activities Commission. He is a Fellow of the AMS. Dr. Colman received his Sc.D. from the Massachusetts Institute of Technology.

Paul J. DeLannoy is the director of natural resource sector services at the Meteorological Service of Canada, Environment Canada. His 30 years of experience in meteorology includes forecasting, teaching, project work, and middle and senior management across all regions of Canada. Mr. DeLannoy is a former manager of the Ottawa Regional Weather Centre of Environment Canada. His specialization for the last eight years has been in the application of meteorology to road and rail transportation. He was on the steering committee for the establishment of the intelligent transportation systems (ITS) Architecture for Canada. He currently is Canada's representative to and English secretary of the World Road Association's Technical Committee 3.3 on Winter Maintenance, and he is a member of the Executive of the Standing International Road Weather Commission. He has written and presented widely throughout Canada, the United States, and in other countries on road weather services. Over the last several years he has championed the establishment in Canada of a national integrated open network of road weather information systems jointly funded by Transport Canada and the provinces and territories. Mr. DeLannoy has a B.S. in physics and mathematics from the University of Winnipeg and a graduate certificate in operational meteorology with Transport Canada.

Francis B. Francois (NAE) is a private consultant and retired executive director, of the American Association of State Highway and Transportation Officials (AASHTO). His field of interest includes transportation systems in all modes, including the planning, financing, construction, and operation of highways and transit, and the conduct of research on transportation problems. Mr. Francois is particularly interested in intelligent transportation systems, including their technology, planning, management and financing, and related institutional issues. Mr. Francois has also been an elected member of the County Council of Prince George's County, Maryland, where he was involved in transportation, public works, environmental, and community development issues. Mr. Francois holds a B.S. in engineering from Iowa State University and a J.D. from George Washington University. He was elected to the National Academy of Engineering in 1999.

George L. Frederick is the general manager of the wind profiler business unit of Vaisala, Inc., in Boulder, Colorado. He received his M.S. in meteorology from the University of Wisconsin, Madison. Mr. Frederick manages a strategic business unit of Vaisala involved with atmospheric projects that include design, installation, and data processing of atmospheric measurement systems employing both in situ and remote-sensing techniques. He works with government, state, and private industry to better employ remote-sensing technology for enhanced monitoring of atmospheric pollutants, aviation safety, and mesoscale weather forecasting. Mr. Frederick is a fellow and past president (1999-2000) of the American Meteorological Society and a former member of the National Research Council's Board on Atmospheric Sciences and Climate.

Frances C. Holt is the chief of the atmospheric research and applications division and director of cooperative research programs at National Environmental Satellite, Data, and Information Service (NESDIS). Mrs. Holt received her B.A. in meteorology from New York University and did her graduate work at the University of Wisconsin. At NESDIS she has been a scientist in the Satellite Applications Laboratory, manager of the Satellite Field Services Station in Anchorage, Alaska, chief of the Environmental Products Group, technical adviser and then acting chief of the User Affairs Office, chief of the Physical Science Branch, and chief of the Atmospheric Research and Applications Division. Mrs. Holt has published over a hundred papers on the use of environmental satellite data. She has also participated in or led many domestic and international training workshops and conferences on satellite applications. She is a member of the American Meteorological Society, a charter member of the National Weather Association, and the founding editor of the *National Weather Digest.* Mrs. Holt is a fellow of the Cooperative Institute for Research in the Atmosphere at Colorado State University and member of numerous program and review boards.

Margaret A. LeMone (NAE) is a senior scientist at the National Center for Atmospheric Research. Her primary scientific interests are the structure and dynamics of the atmosphere's planetary boundary layer and its interaction with the underlying surface and clouds overhead, and the interaction of mesoscale convective with the boundary layer and surface underneath and with the surrounding atmosphere. Dr. LeMone is a fellow of the American Association for the Advancement of Science and of the American Meteorological Society. She is also a member of the National Academy of Engineering and a former member of the Board on Atmospheric Sciences and Climate (BASC). She has served on the National Research Council's BASC Panel on Improving the Effectiveness of U.S. Climate Modeling, and the Special Fields and Interdisciplinary Engineering Peer Committee of the National Academy of Engineering. Dr. LeMone received her Ph.D. in atmospheric sciences from the University of Washington.

Curt Pape is the Road Weather Information System coordinator at the Minnesota Department of Transportation. He coordinates activities associated with the installation and implementation of the statewide Road Weather Information System. This role includes developing procedures and training programs, serving as project manager on research projects, and coordinating program activities with field personnel. Mr. Pape represents the department on several state, national, and international committees and has served as the chair of Aurora, an international group engaged in research, development, and implementation of advanced road and weather technology.

Leland Smithson is the Snow and Ice Cooperative Program coordinator for the American Association of State Highway and Transportation Officials (AASHTO). He recently retired from the Iowa Department of Transportation, where during his 45-year career he was director of various offices in research, maintenance, planning, and administration. Mr. Smithson has extensive highway maintenance experience and has been active in highway maintenance research, serving on the Board of Directors of the Iowa Highway Research Board. He was a founding member and chairman of the Aurora Consortium, an international group engaged in research, development, and implementation of advanced road and weather technology. Mr. Smithson is currently developing a computer-based training program for an anti-icing, road weather information system for the AASHTO Technical Services Program. He chairs the Research Focus Group for the association's Highway Subcommittee on Maintenance. He is a graduate of Iowa State University with a B.S. degree in civil engineering and an M.S. degree in transportation engineering. He is a licensed professional engineer in the State of Iowa.

Richard Wagoner is deputy director and program development manager for the National Center for Atmospheric Research's Research Applications Program. Mr. Wagoner has directed the design and implementation of dozens of weather decision systems around the world, based on advanced science and technology, and has pioneered the system design referred to as Intelligent Weather Systems. He also has been instrumental in diversifying the scope of the Research Application Program's research and development activities by developing new programs in the areas of intelligent transportation systems, agriculture, and the military. He served with the National Weather Service as deputy chief of the Scientific Services Division in the western region headquarters in Salt Lake City, Utah, as area manager for Northern California in San Francisco, and as chief of the Operations Division at the headquarters. Most recently he facilitated the design and implementation of three major programs: the Cooperative Program for Operational Meteorology Education and Training, the Local Data Analysis and Distribution System, and the Federal Aviation Administration's Aviation Weather Research Program. Mr. Wagoner received B.S. and M.S. degrees in meteorology from Texas A&M University. He is a fellow of the American Meteorological Society.

Staff

Amanda Staudt is a program officer with the Board on Atmospheric Sciences and Climate of the National Academies. She received an A.B. in environmental engineering and sciences and a Ph.D. in atmospheric sciences from Harvard University. Her doctorate research involved developing a global three-dimensional chemical transport model to investigate how long-range transport of continental pollutants affects the chemical composition of the remote tropical Pacific troposphere. Since joining the National Academies in 2001, Dr. Staudt has staffed the National Academies review of the U.S. Climate Change Science Program Strategic Plan and the long-standing Climate Research Committee. Dr. Staudt has also worked on studies addressing air quality management in the United States, research priorities for airborne particulate matter, the NARSTO Assessment of the Atmospheric Science on Particulate Matter, and weather forecasting for aviation traffic flow management.

C

Presentations to the Committee

Edward E. Adams, Montana State University: *Pavement Thermal Model Overview*

Bob Baron, Baron Services, Inc.: *WxWorx*

Stanley G. Benjamin, Forecast Systems Laboratory: *Lessons Learned from ACARS for Instrumenting Surface Vehicles*

S. Edward Boselly, Washington State Department of Transportation: *Weather Research Needs: Highway Maintenance*

Ray Derr, Transportation Research Board: *Federal and State Highway Research Programs*

Ian Ferrell, Microsoft Corporation: *Wireless Communication with Drivers*

Robert Hallowell, MIT Lincoln Laboratories: *Winter Weather Road Condition and Treatment*

Jonathan M. Hankey, Virginia Tech Transportation Institute: *Driving in Inclement Weather*

Eldon L. Jacobson, Washington State Department of Transportation: *511 Introduction*

Henry Lieu, Federal Highway Administration: *Weather and Traffic Simulation Models*

Jonathan Lister, Vaisala: *Roadway Sensor Technology*

Clifford F. Mass, University of Washington: *Delivering High-Resolution Weather Information for Washington State Roadways*

Ronald Miller, Ford Motor Company: *In-Vehicle Telematics*

Joseph Perrin, Jr., University of Utah: *Signal Timing during Inclement Weather*

Paul Pisano, Federal Highway Administration: *Overview of the Road Weather Management Program*

Shelley J. Row, Federal Highway Administration: *Surface Transportation and Weather*

Larry Senn, Washington State Department of Transportation: *WSDOT Weather and Traffic Information*

Harold Smith, Iowa State University: *Impact of Weather on Pavement*

Timothy Spangler, COMET: *Cooperative Program for Operational Meteorology, Education and Training*

Richard C. Steed, University of Washington: *Delivering High-Resolution Weather Information for Washington State Roadways*

Andrew D. Stern, Mitretek Systems, Inc.: *The Maintenance Decision Support System: Project Overview*

Jielun Sun, National Center for Atmospheric Research: *Boundary Layer Meteorology*

Jerry R. Waldman, Surface Systems, Inc.: *Transportation Weather and ITS Integration*

Samuel P. Williamson, Federal Coordinator for Meteorological Services and Supporting Research: *Weather Information for Surface Transportation*

Chester G. Wilmot, Louisiana State University: *Emergency Evacuation*